The Earth Manual

The Earth Manual

Malcolm Margolin

DRAWINGS BY MICHAEL HARNEY

A San Francisco Book Company / Houghton Mifflin Book

HOUGHTON MIFFLIN COMPANY 1975 BOSTON

Special thanks to *Huey Johnson*, *Dick Raymond*, and *Life Forum* for the grant that made writing this book possible.

First Printing

A portion of this book has appeared in *Organic Gardening* magazine.

This SAN FRANCISCO BOOK COMPANY/HOUGHTON MIFFLIN BOOK originated in San Francisco and was produced and published jointly. Distribution is by Houghton Mifflin Company, 2 Park Street, Boston, Massachusetts 02107.

This edition published in association with Word Wheel Books, Inc.

 # Contents

Foreword

Between well-trimmed suburban lawns and the vast regions of mountain wilderness, there are millions of patches of land that are semiwild. They may be wood lots, small forests, parks, a farm's "back forty," or even an untended corner of a big back yard — land touched by civilization but far from conquered. This book is about how to take care of such land: how to stop its erosion, heal its scars, cure its injured trees, increase its wildlife, restock it with shrubs and wildflowers, and otherwise work with (rather than against) the wildness of the land.

Wildness comes in many forms. We recognize it easily in mountain lions and thunderstorms. But there is also wildness in a milkweed seed as it bounces lightly along the ground, in a bee twisting and probing within the womb of a wildflower, in the sunlight as it shatters upon the tree tops and pours down through the branches. Wildness cannot be programmed or created; it can only be accepted, and perhaps gently encouraged. Nothing more.

No matter how many acres you own, no matter how noble your ideas may be, you are not a "master planner" with control over the destiny of your wild land. In truth, you're not even a middle-management executive. If you want to encourage the wildness of your land, the very best you can hope to be is a good-natured, easygoing handyman who putters around, solves minor problems, and does a few obviously worthwhile acts.

"But surely you must have an overall plan to guide you," people used to insist. So I'd lead them off into the woods.

"This is where the plans are being made," I'd say. "Scattered, scattered everywhere. In millions of leaves, in billions of seeds, in bird songs thrown down like streamers. In fact, everything in the universe is working together, at every moment and in every action planning out the destiny of these woods."

"It sounds like a complicated plan," people would tell me.

And I'd agree. It is complicated. Too many people ignore the complexities and rush off into wild land with their hands full of tools and their heads full of notions. They'd be better off giving their hands and heads a rest and using their bottoms. For sitting. Try it. Sit long, sit hard, sit in enough different places, and you will develop a sure feel for your land. Bits and pieces of the plan will become clear to you. It may take a long time. But once you develop a feel for where your land is heading, this, more than anything else, will give you the knowledge you need to work in harmony with it.

I have sat long and hard on many fine pieces of land throughout the country. But most of my working experience with land came during my three years at Redwood Regional Park in the hills above Oakland, California. My job consisted largely of running the park's conservation program, and I involved thousands of kids from the San Francisco Bay Area in our various projects. We built erosion-control check dams, planted several thousand trees, and constructed ponds and watering holes. We collected, treated, and dispersed hundreds of bags of wildflower seeds. We restored several acres of meadow land, blended eroding firebreaks into the landscape, nursed injured trees back to health, cut hundreds of yards of trail, logged off part of a fire-gutted eucalyptus forest, stabilized a huge gully and slide area, and did all sorts of things to encourage wildlife.

This book is about what we did, how we did it, and how you can do it too — without money, fancy tools, or fancy skills, in fact, without even any kids. These are all projects you can do on wild land by yourself, or sometimes with the help of a friend or two.

I don't want to pretend that working on wild land is always a picnic. But I expect you'll find it a lot easier than you think. After all, as long as you are working to bring out the wildness of your land, your land will be more than willing to cooperate.

MALCOLM MARGOLIN

Berkeley, California
January 1975

The Earth Manual

1 *Wildlife*

When kids come into the woods, the first thing they want to know is, "Where are all the animals?"

They expect a bear behind every bush, and when after ten minutes they don't even see a rabbit, they feel cheated. "Hey, man, these ain't the *real* woods. All the animals have been killed here."

Before beginning any wildlife project with kids, I found it necessary to prove to them that wildlife does indeed exist and that these are the *real* woods. Yet it's almost impossible to produce a deer when you want one — especially with a group of jumping, yelling, stomping kids. So eventually I learned how to give an exciting animal walk without animals. The idea is to play Indian (in other words, be aware) and look for animal signs. If you've never done this in earnest, you'll be delighted. Try it. Animal signs are everywhere, once you learn to focus on them. There's browsed grass and leaves. There are tracks in the mud. There are deer and rabbit pathways through the meadows as well as mouse tunnels, gopher holes, and mole mounds. There are animal droppings — a sure winner with kids since they combine two of their favorite obsessions: animals and turds. A pile of feathers shows where a bird was eaten. The litter of half-eaten nuts teaches us that animals are enormously sloppy, inefficient feeders — a fact that small children are always glad to hear.

3

Holes in trees and dens in the ground show where animals live. Elsewhere are trees that have been girdled by mice and rabbits, or whose bark has been scraped away by deer rubbing off their velvet. Pellets give away an owl's roost. Hawks and vultures overhead are interesting in themselves and also indicators of lots of life down below.

Another good idea for a wildlife walk is to turn over stones and logs looking for salamanders, toads, mice, and especially snakes. Make sure you replace the rocks to restore these animals' homes. If you're dealing with kids, you might even want to keep a small snake under your shirt (in a cloth bag with a drawstring) to make sure that they'll have a real live snake to handle. Kids don't need to be lectured on the ecological significance of snakes, nor do they need to be told that snakes can't roll up into hoops, as much as they need the experience of holding a real, live, slithering snake.

Or you might try stalking your wildlife like a hunter without a weapon. You can walk a trail for years without ever seeing a fox. Yet make a sincere, strenuous effort to spot one; watch for tracks, droppings, and pathways through the brush; learn everything you can about their habits, hours of activity, and territorial instincts; devote a whole day or more to the task of finding out your local fox; and I guarantee that you will succeed. Not only that, but once you get acquainted with one fox, for some mysterious reason you begin noticing other foxes wherever you go.

However I did it, I felt it was very important to convince kids that the woods were full of wildlife. If at the same time I could convince them that *we* were also wildlife, if I could get them to forage wild foods and think about what it would be like to live as an animal in the wilds, so much the better.

By doing wildlife projects, you get in closer touch with wild animals and at the same time in closer touch with yourself. Kids seem to take this closeness for granted. They are vitally interested in wildlife from their earliest years. They grow up on animal stories and animal crackers. Feeding ducks and pigeons are big events. They clutch teddy bears at night while worrying about the lions outside their windows. Kids love animals, they are terrified of animals, they want to hold animals, they want to help animals, they want to be animals, they are intrigued by the sameness between people and animals, and they are equally intrigued by the wonderful otherness too.

But as we grow, most of us have no contact with real wild animals — only with domesticated animals, stuffed animals, zoo animals, cartoon animals, and advertising animals ("Tony the Tiger sez . . ."). Our emotional and fantasy lives quietly mourn the loss.

Another handicap we bear is the weight of "common knowledge" —the mythology that tells us, for example, that foxes, wolves, ferrets, owls, hawks, eagles, and snakes are vicious, bloodthirsty creatures. Anyone who has dealt even in passing with snakes, owls, and hawks knows how unaggressive these creatures can be, while foxes, wolves, and ferrets can be downright charming and affectionate. I wish that wildlife could form an "Animal Lib" group to "tell it like it is." I think they would point up our terrible and profound fear, ignorance, and exploitation of wild creatures.

Wildlife projects will give you the chance to get reacquainted with animals, to deal with them as they really are, to consider their needs, to experience their vitality, and perhaps to rekindle in yourself a sheer childlike delight in their very existence. It would be unspeakably sad and lonely if we were the only animals on the globe. And it's so much fun to have all these other odd creatures along for the ride.

WILDLIFE CONCEPTS

Before jumping into specific things you can do to foster a healthy, balanced, self-sufficient wildlife population, there are some elementary concepts you should know. These are general wildlife concepts that you should think about whenever you make *any* changes on your land at all.

Wildlife needs   Wildlife needs food, cover, and water. It needs all these things within a fairly compact area — how compact depends on the size of an animal's "territory." If any one of these things is missing, or in short supply, this will limit the number of creatures the land can support.

The next time you look at a piece of land, chant over and over again, like a mantra,

> Food, cover, water;
> Food, cover, water;
> Food, cover, water.

It will help you see the land as an animal might see it. A thick forest may have plenty of cover but no food. An irrigated farm may have plenty of water and enough food to feast Noah's Ark, but if there is no cover, it will be virtually a wildlife desert.

Water is sometimes difficult to judge. If it rains throughout the year, or if there is generally fog or heavy dew, water may not be a major problem. If you think you have a water problem, see pages 150-64 for how to build ponds and watering holes.

Law of the minimum   Averages mean a lot to insurance brokers and baseball players, but they mean absolutely nothing to wildlife. It is not the average condition but the most extreme condition that determines wildlife population. Your land may have enough average rainfall for a sizable quail population. But if every year, or even every two or three years, there is a month-long drought, it is this extreme condition of drought that will determine how many quail can survive. Likewise, all wildlife has to adapt itself to the extremes of food shortage, temperature, water shortage, or the elimination of cover. If you want to help your wildlife, the best areas to focus your attention are at the various extremes — a periodic food shortage in March, for example, a drought in August, or a severe freeze in February might be good places to begin.

Do not restock   Do not worry about restocking on thinly populated land — unless, of course, the animal you're interested in has been totally exterminated from your area. If there is even a remnant, no matter how skinny, suppressed, and discouraged, it will do the restocking for you.

Instead of restocking, what you should be doing is improving the habitat and thus the *carrying capacity* of the land. On good grouse habitat, for example, you can find perhaps one grouse every 3 or 4 acres. On bad habitat you'll be lucky to find one in 200 acres. If you release more grouse in a bad grouse habitat, they'll die off very

quickly — within a few weeks even — until their numbers are back to what the habitat can support.

Once you improve the habitat, however, you can take advantage of wildlife population dynamics — a fantastically explosive force. You've undoubtedly read those figures for rabbits or mice — showing how if all the bunnies from one pair of rabbits lived and reproduced, and if all their offspring lived and reproduced, within a few years we'd need periscopes to see each other. It's true for all wildlife.

If the habitat's right, your animals will have far more fun re-stocking the land than you will — and they'll do a much better job of it too.

Predators   I'm sure it's not necessary to tell you that predators do not wipe out their prey. In national parks where both pred-ators and prey are equally protected, I have never heard of a single case where the existence of the prey was endangered. In fact, despite healthy predation, the most common problem in such protected areas is still overpopulation of deer, rabbits, and other prey animals.

Variety   There is no such thing as the perfect habitat for all wildlife. Every species of animal has its own preferences. If you strip a piece of land of all its vegetation and let it alone, it will even-tually go through several stages of weeds, grasses, shrubs, small trees, and big trees until it reaches its climax. At each stage it will be a preferred habitat of different animals. Animals that like fields will be evicted by the eventual arrival of brush. Animals that like shrubs will be made homeless by the forest. If you want a variety of wildlife, the thing to do is to create as many varieties of habitat as you can within the bounds of your natural environment.

When two environments meet, by the way, the wildlife possibilities are multiplied many times over. This is known as the *edge effect*. The edge of a forest is far more fruitful than the center. Other exciting places are the shores of lakes and ponds, the borders of meadow land and brush, and (for birds) the billowy area where the tree canopy meets the sky.

Manage your land for all wildlife It is important to keep *all* your wildlife in mind, especially when you are dealing with experts who think that *wildlife* means *game animals*. Take, for example, the matter of prescribed burns, where you burn off the forest floor under controlled conditions. Prescribed burns are very fashionable these days. It's the "in" thing to treat Smokey with condescension and burn off a few acres every year. The idea behind burning, as far as wildlife management goes, is that on the charred soil a lush crop of weeds, berries, legumes, and grasses will grow up which are very good for deer, grouse, moose, quail, bear, and various other game animals. This is true. But remember that the same fire is also going to destroy millions of mice, rats, ground squirrels, lemmings, worms, insects, soil organisms, snakes, toads, newts, rodents, amphibians, reptiles, and the predators that depend on them. I'm not saying you should *never* burn, but if you do, keep in mind *all* your wildlife.

BRUSH PILES

Architecturally, there is not much to building a brush pile. You simply pile up a lot of brush with the idea of providing shelter for small animals. If you want to get more elaborate, you can put heavy logs or rocks on the bottom to prevent the pile from matting down to the ground, thus keeping it open for the entrances and exits of the animals you attract. You should pile the brush as high as you can — at least four or five feet high — and in general try to keep it twice as wide as it is high. If you have some more heavy logs or limbs, throw them on top to compress the pile and keep it from blowing apart, and, *voilà*, you have the perfect brush pile. Members of the American Institute of Architects might blanch when they see it — but they tend to be snobbish anyway. Your wildlife will love it. A brush pile is nature's very own idea of style, borrowed from the dwellings of muskrats, pack rats, and beavers. It's also a wonderful project for a group of little kids.

Where to get brush I would not cut down brush especially to make a brush pile, but you'll usually have lots of material available: dead trees; slash from logging, thinning, pruning, or trail building; old fence posts; or brush from a brush-clearance (meadow-restoration) project. Another good source of brush is a collection of your community's old Christmas trees — a good way of recycling Christmas trees and involving your community in your land.

Where to build brush piles The best place for a brush pile is where there is lots of food, lots of water, but no shelter. Meadows and clear forests are good places. Remember that the animals you attract will be small and not very far ranging, so that the closer they are to food and water, the better they will survive.

The future of a brush pile I've always admired the old-car method of berry farming. It works like this. Just drive an old car into a field, desert it, and within a few years it will be totally covered with berries — with no planting, no cultivating, no watering on your part.

The old-car method works equally well for brush piles. You can plant cuttings if you want, but even if you do absolutely nothing, the brush pile will soon be taken over by blackberries, Virginia creeper, bittersweet, wild grapes, and other berries. The seeds are brought by birds who perch on the brush, and, while adding songs to the air, gently perk up their tails and drop a few seeds into the brush pile. The brush itself decays into humus with time, but the big mound of brambles and berries is likely to remain for many years.

Fish brush piles Brush piles work just as beautifully for fish as they do for birds and animals. A brush pile in the water gives small fish a place to escape to, gives fry a place to hatch and grow up, and gives algae a place to cling to and produce food.

The problem with a fish brush pile, of course, is that brush can float away. But if you can find a still backwater or the edge of a small pond, throw in some brush, stake it or weigh it down, and it will serve your fish population very nicely.

The only thing to watch out for in building a fish brush pile is overenthusiasm. Don't overdo it if you are dealing with an enclosed body of water like a pond or a small lake. A moderate amount of cover will benefit your fish. Too much cover, however, and the fish will overpopulate. Unlike mammals, overpopulations of fish do not usually starve; rather, horror of horrors, they dwarf. Instead of a few dozen healthy trout, you end up with hundreds, even thousands, of trout ignominiously going through life as minnows. So, unless you're into miniaturization, I suggest you go easy on fish brush piles.

Rock piles Piles of rock are not quite as effective as brush piles, but if you are in a field with lots of rock and no brush, you might pile some of the rocks anyway. Here again, make the pile as big as

you can, leave some space at the bottom for entrances and exits, and you will very likely provide a fine home for some animal. Or for some fish if you pile the stones in the water.

Living brush piles   Living brush piles is not the name of a new rock group. It is a way of getting your vegetation to look and act like brush piles.

One good way of doing this is to cut halfway through a tree and let it fall with the lean. (See page 34.) This way it will stay alive for a few seasons, providing browse and shelter for many animals.

Another trick you might consider is multiple layering. You can convert a fairly modest bush into a huge, jumbly mass of thickets by staking some of its branches underground. (For detailed instructions on how to layer, see pages 108-10.)

The inhabitants   The animals that live in brush piles are generally secretive. Yet if you place the pile in a good location, you'll be surprised at the activity. You may never see an animal, but tracks, well-worn paths, droppings, songs of birds, peepings and whistlings, the interest of a hawk overhead, and scamperings and rustlings every time you pass are all indications of brush-pile prosperity.

HOLES AND DENS

Animals simply love to live in holes. They love caves, holes in trees, and dens in the ground. Not long ago humans were among the animals competing for big, clean holes and caves. And what high standards we still have! Nothing worse can be said of a dwelling than "it's an absolute hole." Yet people who can afford it make a separate room which they call a "den," panel it with wood, keep it warm, stuffy, natural, and holelike, and show it off to their friends. If you decide to make a few animal holes around your land, it would be wise to remember this human situation. Not all holes are created equal. A nasty, ticky-tacky hole might do in an emergency, but a good hole is an animal's idea of true wealth. Spend some time making the hole comfortable, well drained, and secure, and it will attract an animal and make it happy for many years.

Den trees   In old forests, you will find lots of dying trees, lots of insects to burrow into the dying trees, and lots of woodpeckers to eat the insects. In the process, woodpeckers make holes for their own nests, and these holes later become dens for owls, squirrels, several kinds of birds, possums, raccoons, and other creatures.

Without holes to nest in, squirrels will build leaf nests. One look at these clumsy, bulky, inept-looking structures will let you know that the squirrels would much rather have a den. Also, because of exposure to the weather and to predators, squirrels are rarely successful in raising a family in a leaf nest.

One of the best and most satisfying projects I know is to clear out rot from an old tree injury, creating a well-drained cavity in the process. Not only will this keep the tree sound, it will also provide a premium den site. (See pages 52-54.)

Another thing you can do, I suppose, is build wooden boxes, which most definitely attract wood ducks, squirrels, and other animals. If you want to create a sort of shantytown for wood ducks, there are instructions and plans available elsewhere. I find the idea aesthetically silly and condescending toward animals — but the animals don't seem to mind very much, so maybe I'm just being a purist.

One thing you might consider, especially if you have a predominantly second-growth forest of adolescent trees, is to create a dead-tree environment that will attract insects that will, in turn, attract woodpeckers. Unfortunately, there is only one way to create a dead-tree environment: you have to go out and kill trees. This makes for a rather heavy — although occasionally necessary — conservation project. The best way of going about it is to find a few larger trees and girdle them. Use a chisel and mallet to cut a wide band around the tree. The band should be about two inches wide and deep enough to remove all the bark and cambium (the green juicy layer just under the bark). The tree will die on its feet and will hopefully remain standing for several years, attracting termites, insects, woodpeckers, and the other lively guests that attend the dying of a tree.

Ground dens   What a woodpecker does in the trees, ground hogs do on the land. Ground hogs (also called woodchucks or marmots) are the housing engineers and contractors of the woods, building roomy dens that are later taken over by foxes, skunks, rabbits, snakes, raccoons, possums, and other creatures.

If your land is short of holes, one of the best long-range measures is to encourage ground hogs — as long as you don't encourage them too close to any garden or nursery you're planning to build. Unfortunately, the human race has been so eager to get rid of ground hogs that no one has done any research on how to attract them. I once asked an old-timer how you go about attracting ground hogs, and all I got was one of those long, sarcastic, head-scratching, "what

is this world coming to?" sort of stares. The best way I know is to build a few brush piles, under which ground hogs dearly love to build their dens.

Until such time as you can transfer conservation duties to the ground hogs, you might want to build some dens yourself. Build them near cover, near winter food, and near water, and make sure they are well drained so that you don't drown the animal you will attract.

A good den can be made out of old pipes, drain tiles, or sections of culvert. Lay them down in some inconspicuous place and heap stones, dirt, and brush over them — partly to disguise them, partly to keep the sun from baking the inhabitants. Make sure there are at least two exits and that all the entrances and exits are well disguised with brush or stone.

Another good den can be made from rocks and a piece of old plywood or sheet metal. Simply arrange the rocks on the ground and put the wood or the metal on top to form a roof. Cover everything with lots of earth and rocks, and again remember to provide cover at the entrances and exits.

Building a den is easy. Now comes the hard part. You've got to stay away. Don't keep returning to shine flashlights into the holes or beat on the den with your walking stick. Do the best thing any landlord can do — stay out of your tenants' way. Eventually you'll notice a well-worn path leading up to the entrance. This may be your only reward for a job well done.

VARIATIONS ON A HEDGEROW

If you have any open meadows, you should definitely consider putting in a few scraggly hedgerows. Meadows and fields have lots of food for wildlife, but lack of cover often keeps animals away. No rabbit in his right mind wants to get caught in the middle of a field.

A hedgerow is a row of thick, bushy plants which provides both shelter and a passageway for animals to get across the field. Any

Culvert den

bushy plant will make a good hedgerow, but the best type of plant is one that both is thorny and bears an edible fruit.

How to make a wild hedgerow   If you decide to break up a field with a hedgerow, you should try to avoid a straight line of bushes. It looks artificial. You can learn how to plant a wild-looking hedgerow by watching an animal, even a cat, make its way across a field. It will seldom make a beeline but will run from a bush to a rock, sidle along a fallen log, dash under a fence rail, and zigzag from one clump of weeds to another. In planning your hedgerow, keep this image in mind. There are probably several good sheltered places already in your field. All you have to do is connect them with a brush pile, a log, a pile of stones, and a few shrubs that you can plant. Kids, by the way, really enjoy planning out a hedgerow of this type. It's like playing hide-and-seek.

What to plant   In deciding what to plant, keep your eyes open to see what in your neighborhood grows close to the ground in a thick, hedgelike way. You can sometimes get good advice from your local Fish and Game Department, but don't let them sell you on a foreign exotic like lespedeza or multiflora rose (which spreads like crazy), and beware of anything that needs constant pruning to keep it shrubby.

In California there are various shrubs, especially ceanothus or buckthorn, that make good cover and provide some food in the bargain. In other parts of the country, hedgerow materials with a fairly good reputation are Osage orange, high-bush cranberry (*Viburnum trilobum*), bayberry, and various small dogwoods. Spruce is especially good, since it grows close to the ground, has sharp needles that discourage larger predators, and doesn't lose its leaves in the winter.

There are various ways of planting shrubs. You can grow them from seed and plant them out as you would a tree. Or you can take cuttings and, after treating them properly, plant them out as a hedgerow (see pages 104-18).

Ditches   One place where you can apply the hedgerow concept of cover and passageway is along drainage ditches and gullies, wherever they penetrate into a field. Ditches and gullies often have plenty of water. By planting berries, shrubs, and vines, you can add cover and food, thereby completing the formula for wildlife intensity. And by dense planting you may be able to prevent further erosion along drainage ditches.

Predators   Making hedgerows and planting ditches will help various small animals outsmart their predators. But don't feel sorry for your local foxes and hawks. Any improvement in the health and population of a prey animal is also an improvement for the predator. In fact, if you are one of those sainted, wild souls who manages land especially for eagles, wolves, foxes, or other predators, making hedgerows is one of the best things you can do. Whereas without hedgerows your field can support perhaps only two families of rabbits, it can now support four. These extra two families will produce more than forty bunnies a year, which will make the predators very happy indeed.

Corridors   Another good project to keep in mind is creating the mirror image of hedgerows — namely, corridors through heavy brush. The idea here is to clear out strips to allow animals to penetrate the brush land for browse. Deer, in particular, will benefit. Avoid straight lines and save yourself work by connecting already existing clearings. Also, don't be too concerned with pulling the brush out by the roots. If the corridors are used regularly, any new growth will be trampled or browsed.

DOGS AND CATS

Dogs and cats are a joy in the living room or in the back yard, but in the wilds they are a first-class problem. Wild lands are full of them, some of them escaped pets but most deserted by owners who felt the time had come for their tabbies to fend for themselves.

In the hills above Oakland we had dozens of tabbies fending quite nicely for themselves. We also had a pack of wild dogs led by an absurdly civilized poodle. The dogs ran down an occasional deer, but they didn't bother me half as much as the several dozen cats who lived in the park and were doing incredible damage to the rodents, reptiles, and especially to the quail and other birds. The cats would wipe out nest after nest, even eating the eggs. They killed more than they needed, playing with what they couldn't eat. Once I even saw a cat carry off a huge kicking squirrel at least its own size. To give these animals credit, it has been argued that they were merely filling a predator's niche. That may be true, but I still would rather have had a population of bobcats, foxes, coyotes, and wolves instead of the escaped dogs and cats who were edging them out.

But whatever my feelings may have been, my efforts to eliminate the cats were dismal — even farcical. Once a couple of older kids and I borrowed a few Havahart traps, and we began to trap cats. Trapping them was no problem. Any smelly cat food worked. Sometimes we caught wood rats, skunks, raccoons, or foxes, but at least three out of four times we caught a cat. *That* was the problem. The kids would bring the cat to me, and I'd do the dirty work of drowning or shooting it. Every day I found myself wishing the kids would get discouraged and quit. But they didn't. They kept coming back with more cats.

A few ladies who lived around the outskirts of the park heard about the project and were totally horrified. I became known as an ogre and cat killer. One of the ladies was so upset that she brought a pile of blankets and a few cans of cat food and offered to adopt any of the little darlings I caught. Each afternoon I would look at the snarling, scar-faced, mangy, vicious creatures the kids had brought in, but none of them seemed to qualify, even remotely, as a "little darling."

In any case, after a few weeks we had caught lots of cats, but we still had not seriously dented the feral cat population. And I was thoroughly sick of the whole business. There is *no* aesthetically pleasing way of killing a cat, and I was thoroughly repelled by having to do this ugly act over and over again. So I eventually copped out on the bobcats, the foxes, the coyotes, and all the fledgling birds, mice, and squirrels. I still think that getting rid of wild cats and dogs is an extremely worthwhile conservation project for someone else to do, but, no thanks, not for me.

CLEARINGS

Some people find it hard to believe that by cutting down trees you can greatly increase your wildlife. Yet it is definitely true. Dense forests are very poor wildlife habitats. Create a few openings, however, and you will encourage a much more varied and plentiful wildlife population.

We usually think of clearings as man-made, but they are very much part of the natural landscape. They are caused by flooding, fire, windstorm, and certain conditions of soil, rainfall, and exposure. What makes them so good for wildlife is that the sun reaches the ground, causing a lush growth of weeds, berry bushes, shrubs, grasses, and fruit trees that would otherwise be shaded out in a forest. Trees at the edge of openings also receive lots of sun, which allows them to spread out and keep their lower branches — thus providing easy-to-get-at food and shelter for many animals.

Many Indians understood the value of openings in the woods, and throughout the country they were reported to have burned the land periodically to maintain them.

The best time to think about openings is when you're planting trees. Do not plant a thick, unbroken forest. Leave small half-acre or one-acre clearings here and there — say, one every quarter mile or so. If you make lots of small clearings, you will touch the borders of many home territories and benefit the maximum number of animals.

Another good project is to keep track of your current openings and visit them every year or two with lopping shears and a grub hoe to keep the forest trees from gaining a foothold.

If you are going to clear an opening out of an already established forest, you'd be best off with a sunny exposure (either south or west). You can pile the slash into brush piles. If you want the opening to remain open, be sure to watch the stumps, lest they resprout (see page 35). Another consideration: you might do the felling during the critical food period of late winter and early spring. The buds and twigs of the trees you chop down — especially maple and birch — will provide good emergency browse.

Dusting places   After you make a clearing, a very nice amenity to provide is a place where animals can dust off. Many birds and small mammals are made very happy by dust baths — perhaps it removes parasites, perhaps it serves some other function, or perhaps (a reason orthodox zoologists find hard to accept) it's just a whole lot of fun. Whatever its purpose, you might want to throw a few handfuls of sand, dry earth, crumbly wood, or sawdust into the middle of a clearing, where it will be well used and appreciated.

Forested openings   Some forest trees tend to darken the ground and eliminate other vegetation. But there are other trees — especially locusts, alders, and aspen — that actually encourage growth beneath them. These trees give a light shade (rather than a heavy shade) and, more important, they encourage nitrification — that is, they foster the bacteria that convert atmospheric nitrogen into nitrogen fertilizer. Not only is the ground beneath these trees full of shrubs, bushes, weeds, vines, and herbs, but these plants are actually more nutritious than the same plants growing somewhere else. You might do well to plant these trees and protect them wherever they are now growing, especially since the lumber industry considers them "weed trees" and on commercial land is doing everything it can to replace them with money-making conifers.

FOOD

Planting food for wildlife is a delightful switch. For ages wild animals have been used to feed humans; now you have the chance to feed them instead.

Before getting into specific suggestions, let's examine the yearly food cycle of, say, a deer.

In the late spring and throughout the summer, there is plenty of green grass, weeds, wildflowers, berries, and fruit. "Livin' is easy," as the song about summertime goes.

In the fall, grains and many seeds ripen, but the big event is the nut harvest. Acorns, beechnuts, hickories, hazelnuts — all fantastically nutritious — are readily available until the first snows. The deer put on a good layer of fat, and they are ready for winter.

In the winter, deer paw through the snow after acorns and eat whatever tough greens they can find (especially wintergreen). By the end of winter, however, they are reduced to eating buds and twigs. They seem to go about it with all the gusto of a starving man eating his boots. Buds and twigs are definitely a last-resort food.

At the beginning of spring (about late February through the first week of April), there is a general food emergency in the woods. Deer and other animals have used up their layers of fat. The ground has been scoured for acorns and nuts. The browse has been pretty well eaten, and the earliest leafy greens have not yet come. If you want to increase your wildlife population significantly, this is the season when you should concentrate your efforts.

What exactly should you plant? It varies from one area to the next. Your local Fish and Game Bureau is usually a good source of specific suggestions. But in general, here are some things you ought to be thinking about.

Nut trees Planting nut trees, or *mast*, is probably the best thing you can do for your wildlife. It builds them up in the fall and keeps them alive through the winter. Oaks, beeches, hickories, black walnuts, butternuts, hazelnuts, chinquapin, piñon pines, and junipers are among the most valuable nut trees.

You can plant nut tree seedlings or you can collect nuts in the fall and plant them directly in the field. However you do it, you should plant as wide a variety as your environment can support. Do not pin all your hopes on just one species of tree, since nut trees are enormously unreliable about producing. For five years a tree may be bountiful, then one year it will seem to forget. If you want to assure your animals of a steady diet of nuts, you'll have to plant several different species.

A word about acorns, which are singly the most valuable wildlife mast. Some oaks, as you may know, produce sweet acorns. These are the so-called "white" oaks. The "red" oaks, on the other hand, produce such bitter acorns, so full of tannin, that just looking at them puckers you up right down to your toes. Despite popular opinion, however, the bitter acorns are extremely valuable to wildlife. In the fall animals prefer sweet acorns, which play a major role in fattening them up. But sweet acorns rot quickly. The tannin in the bitter acorns prevents them from rotting, keeping them sound often until the next spring.

Release cutting   Release cutting is a very valuable aid to nut and fruit trees, most of which are transitional. It is their destiny to be overshadowed. You and your trusty ax can overcome destiny, however. If you clear around an especially valuable nut or fruit tree, you will "release" it and reap double benefits. The tree will spread out and bear more heavily, while underneath it shrubs and greens will spring up for wildlife ecstasy.

Freeze-ripened fruits   There are certain fruits that stay on the tree throughout the winter. In the early winter they are often too bitter or otherwise unpalatable. But by late winter the repeated freezing and thawing have made them quite edible — just at the season when they are most needed. Among the plants that generally follow this pattern are Oregon grape, crab apples, mountain ash, high-bush cranberry, hawthorn, staghorn sumac, winterberry, coral-berry, and partridgeberry.

Early foods Another good thing to do is to plant a variety of the earliest foods, foods that are ready to be eaten just at the end of winter. Elms generally have seeds that ripen very early, and so do other plants. Consult your local Fish and Game people, your Soil Conservation District, or (the ultimate revolution!) your own eyes for information about the earliest-ripening forage plants in your area.

Gourmet diets Even during the spring, summer, and fall, certain animals have very narrow food preferences. Birds, in particular, are often very fussy about what they'll eat. There is no way you can attract acorn woodpeckers without acorns. Here again, you might get local advice from the Fish and Game Bureau or from your local chapter of the Audubon Society about what other foods you might plant — especially in conjunction with reforestation projects, erosion-control plantings, or other such projects.

Artificial feeding As a matter of theory, I'd avoid artificial feeding of wildlife. In an overbrowsed area, feeding deer will merely aggravate the problem. Feeding waterfowl may convince them to winter further north than usual — with potentially dangerous results. In fact, wildlife theory is now swinging to a hard line that you should *never* give handouts to your wildlife. So much for the theory. When it gets cold and nasty and when the animals you love are obviously having trouble, and you just happen to be sitting on a bale of alfalfa — well, I'll leave you to cope with the "theory" as best you can.

If you decide not to resort to artificial feeding as a regular thing, there are still several abnormal circumstances under which you might consider it:

For rare, endangered, or threatened species.

During extreme emergencies. Not necessarily the coldest week of the year, but certainly during the coldest week in fifty years.

When some act of civilization has deprived your wildlife of its usual food. A logging operation may have cleared out most of the nut trees and replanted the area with nothing but pine. Or perhaps a housing development has gone up in a sheltered valley where elk would regularly congregate during the coldest part of the winter. Under circumstances like these, you can feel perfectly justified in putting out some food — but once you begin, please keep it up on a regular basis.

The best emergency foods for small animals and birds are corn or chicken feed. For larger animals, corn, hay, or alfalfa will be superwelcome.

READING

Here are some of the books I have found helpful.

The Way to Game Abundance, by Wallace Byron Grange. New York: Scribner's, 1949.

This is an absorbing book dealing with the larger issues of cycles, population mathematics, predation, territoriality, and vegetation succession. The issues are all brilliantly covered in intelligent, readable prose. It's a book that examines all the clichés — their clichés ("All predators are bad") and our clichés ("All predators are good") in light of the author's experience. This is a rare book that not only covers the bigger issues but does so in such a manner that you can use the information in a practical way.

The book is also fun to read.

Wildlife Habitat Improvement Handbook, Forest Service Handbook FSH 2609.11. Washington, D.C.: U.S. Forest Service, 1969.

Written for Forest Service personnel, this handbook has good, sound, practical information and advice. It has especially good chapters on stream improvement (for fish) and on water-hole construction. It also has good lists of plants and trees for waterfowl.

The Farmer and Wildlife, by Durward L. Allen. Second revised edition. Washington, D.C.: Wildlife Management Institute, 1970.

This booklet is aimed at the farmer and his concerns: cultivation, irrigation, and grazing. The book focuses on how to reconcile the farmer's interests with those of wildlife. It is sixty-two pages long and is available free from the Wildlife Management Institute, Wire Building, Washington, D.C. 20005.

American Wildlife and Plants, by Alexander C. Martin, Herbert S. Zim, and Arnold L. Nelson. New York: McGraw-Hill, 1951.

This is a big book and very thorough. It covers all major species of birds and mammals, one by one, giving the food habits of each. It breaks down these habits by geographical area, season, and the importance any one food plays in the total diet. Next it turns to the plants, listing major species of grass, herbs, aquatic plants, vines, shrubs, and trees, and telling what animals eat them and where.

This extraordinarily valuable book is also available in a Dover paperback edition.

2 *Felling a Tree*

This chapter will tell you how to cut down a tree. Don't be shocked. No matter how much you love trees, there are many times when you will have to fell them. And when you do, you will probably discover a secret that most sensitive conservationists try to hide — even from themselves. It's fun to chop down trees! It's fun to look a tree over, plan its fall, and work hard against a real physical objective. There is a definite thrill when, after twenty minutes of sawing, you hear the telltale crack, there is a long, pregnant pause, and the tree slowly, ever so slowly, begins its final journey to the ground. It's very exciting to yell, "Timber!" and backstep rapidly away, all your senses wildly alert. And there is something overpowering and deeply satisfying when the tree hits the gound with a huge crash, the ground shakes, and then a tremendous, complete silence follows.

As a conservation project, I found tree felling hard to beat. Kids just love it. It doesn't matter if the kids are teen-age boys on a macho trip, a group of girls from a parochial high school, or fifth graders on a bird hike. If there is a tree to be chopped down, kids will fight for the privilege. This used to disturb me. Here I was, a hotshot conservationist, supposedly leading the younger generation off to a new, gentle ecological consciousness — yet our spirit was unmistakably that of the Goths about to sack Rome.

26

I've always been troubled by the spiritual aspects of felling trees, and I don't have any easy solutions. Kids love to chop down trees, and so indeed do I; yet at the same time, I am appalled at the eagerness and heartlessness with which we take on the job. Felling trees is the only project I know where having fun is a problem. I wish I had more of this problem on other projects I've been involved in.

PLANNING THE DOWNFALL

Why fell a tree There are many technical, textbookish reasons for cutting down trees under the guise of "forest management." But most forests need far less managing than professionals pretend. Nevertheless, there are still times when you will want to fell a tree. For example:

EXOTICS You might want to rid your land of some foreign exotic (like eucalyptus in California) where it is ecologically inappropriate and where it is crowding out native vegetation and the life systems that native vegetation supports.

RELEASE CUTTING You might want to remove competition from around an especially valuable nut or fruit-bearing tree if you feel that it is being crowded or overshadowed.

TRAIL BUILDING Most of the time you can work your way around a tree, but there are rare times when you will have to cut.

THINNING Sometimes after logging, burning, or overenthusiastic planting, the land will sprout many thousands of trees growing close together, crowding each other out, and dwarfing each other. Instead of a noble forest, you might find a thicket of toothpicks. Time will take care of the problem, if you're willing to wait a hundred or so years. But if you're impatient, you can thin the forest yourself.

WILDLIFE CUTTINGS If you have a uniformly dense forest, you might want to create a few scattered openings where grasses, weeds, legumes, and shrubs can grow. Or, during a severe winter, you might want to fell a tree here and there to give deer, rabbits, and elk (if you're so lucky) some tender buds and twigs to browse on.

Which trees to cut This will depend largely on why you are cutting them. Let's say you are thinning or making small clearings. You might take your cues from the lumber industry, which divides all trees into roughly two categories: "good" trees (i.e., trees that make money) and "bad" trees (trees that don't make money). Bad trees, needless to say, must be gotten rid of, and if you seek professional advice, you will be urged to chop down "wolf" trees, "weed" trees, trees that are too young, trees that are too old, crooked trees, and so on. The goal of traditional thinning methods is to produce valuable, uniform lumber. Lumber people really get turned on by uniformity, but that is exactly what you do not want. You want as much diversity as you can get within the bounds of your natural environment. So go easy on professional advice and ignore the traditional rules for thinning (if you happen to know them). Remember that a single, half-rotten, malformed oak tree might have greater wildlife value than an acre of thriving young conifers. As you thin, try to create an authentic environment (whatever that means for your area) and a varied one. Add a touch of craziness here and there, and you will end up with a forest with lots of wildlife, lots of different plants, and lots of interesting trees with unloggable personalities.

Erosion Whenever you cut a tree, you should be thinking about erosion. What is going to hold the soil together once the tree is down? Don't cut too much at any one time, especially on slopes. Clear in small patches, leaving trees standing (at least temporarily) to hold the soil and serve as windbreaks.

To further minimize erosion, you might try dropping trees perpendicular to the flow of water. Left in place, they will act as check dams, slowing down the water and perhaps preventing the development of gullies.

Clear a work area Sawing down a tree is hard work, and you must have a comfortable area to do it in. You most certainly do not want to work in cramped quarters, your body contorted like a pretzel. Clear out a place where you can get a firm footing, and if you're using an ax, remove enough brush so that you can get a free swing. Lop off any lower branches of the tree that are in your way. Then clear out an escape route or two, preferably at 135-degree angles to the direction you expect the tree to fall. Fix the escape routes firmly in your mind.

Looking a tree over　Before felling a tree, walk around it and study it. Check it once more for value. Is it a den tree (one with holes)? Is it a fruit tree? A nut tree? Are there any nests? Is it a rare species?

Then look it over for safety. Be sure to avoid trees with dead branches. Dead branches are called "widow makers," for obvious reasons.

Judging the fall　Now think about where the tree will naturally fall. Which way is it leaning? Stand back and hold your ax out before you by the tip of the handle, with the blade pointing at the tree. The handle will be *plumb*, and by sighting along it, you can determine the lean.

Look at the balance. Too many heavy branches on one side may help pull the tree over to that side. Pay careful attention to the top. If the top is nodding in one direction, that's the way the tree is likely to fall.

Next you should check the trunk at about waist height, where you will be cutting. If you see any rot, or if in fact at any time during the cutting you feel yourself sawing through rot, remember that this will influence the fall toward the weaker side.

Finally, take the wind into account, especially if the tree has lots of foliage. Add these four factors together—lean, balance, rot and wind—and weighing the lean most heavily, you will get a good idea of where the tree wants to fall.

Once you have figured out where the tree wants to fall, file that information away and begin thinking about what is best for you. Look for a clear space (or *bed* as it's called) into which you can

drop the tree where it won't damage anything valuable or get hung up in the branches of another tree. But do not, under any circumstances, consider dropping a tree up a steep slope: it is very likely to hit against the slope and kick back at you.

You now know where you want to drop the tree and where the tree wants to fall. Do you both agree? If you and the tree are within forty-five degrees of each other, you can proceed without any special skills or equipment.

Roping Let's say, however, that you and the tree have very different ideas on where it should fall. If you expect to get your way, you'll need a rope or cable. Tie the rope as high on the tree as you can. Then, as you cut, have someone pull hard and constantly in the direction you want the tree to fall. Needless to say, make the rope long enough so that whoever is pulling does not end up with a tree on top of him. Often it is better to run the rope through a pulley or around another tree and then back beyond the tree you are cutting down.

For a smaller tree, you might dispense with the rope and simply have someone push against it with a forked stick, positioning the stick high on the tree for maximum leverage.

DROPPING A TREE

Once you have cleared a work area, determined the direction of the fall, made sure of your escape routes, and perhaps attached a rope or cable, you are ready to cut. In the following pages, I will tell you how to do it. The basic procedures are not very difficult to learn, but please don't consider yourself an expert just because you have read this chapter.

There is no book in the world that is going to teach you how to fell trees with the assurance, grace, and safety of an expert lumberjack. Be prepared for the eventuality that now and then, especially in the beginning, you will mess things up and drop a tree in the wrong direction. When this happens, don't just pass it off as something that *experience* will eventually take care of. *Experience* won't take care of a blessed thing — unless you think deeply and creatively about it as it happens. Learning from mistakes is always a lot harder than moralists pretend — especially when your hands are blistering from an ax, your arms are aching from a hand saw, or perhaps your whole body is dulled from the buzzing and vibrating of a chain saw. Yet if you want to become competent at felling trees, it is under circumstances such as these that you will have to be most alert and thoughtful.

Make a few mistakes, figure out why, and eventually you'll learn the craft. Until then, I hope you'll think of yourself as an apprentice. I especially hope that you'll serve out your apprenticeship deep in the woods — not, for goodness' sake, in your back yard, book in one hand and saw in the other, family and friends crowded around in admiration, while you undertake to drop a 200-foot elm into a narrow space between your house, a neighbor's house, and some high-voltage power lines.

Small trees   For small trees up to about six inches in diameter, you do not need any fancy, formal cuts. An ax will do the job as well as a saw. Chop a notch into the tree on the side on which you want it to fall. Keep chopping. When you are about three-quarters of the way through the tree, you can usually stand off to one side and push it over into its bed. Pushing with a forked stick is particularly effective.

Tools   The traditional way of felling bigger trees is with an undercut and then a back cut. It is possible to use an ax, but it is also dangerous. The chopping tends to dislodge weak branches, which can fall on your head, and the lack of precision in an ax cut means a certain lack of predictability in how the tree falls. Later on I'll tell you how to use an ax, but if at all possible, do it with a saw. There are various big-toothed hand saws that do the job, but my favorite is a bow saw with a tubular frame. A chain saw makes the job a lot easier but a lot more dangerous.

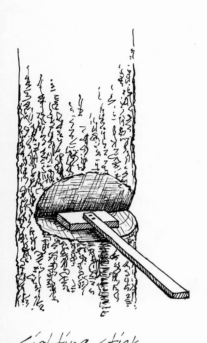

sighting stick

The undercut The undercut is made on the side toward which you want the tree to fall. Begin at about waist height with a horizontal cut about one-third the distance through the tree. Then pull out the saw and begin another cut well above the first cut, angling it down until you have cut out a wedge. The inside angle of the wedge should be at least forty-five degrees.

Before going on to the back cut, check what you have done. Make especially sure that the inside edge of the undercut is perpendicular to the direction of the fall. You can check this with a handmade sighting stick, or with a straight-handled, double-headed ax. Facing the cut, insert the ax head against the inner edge, and the handle will point to the direction of the fall.

Back cut Once you are comfortably certain that the undercut is correct, you can begin the back cut, or *felling cut*. Go around to the other side of the tree, opposite the undercut side, and saw into the tree about two inches above the base of the undercut. Keep the cut level. Don't angle it down.

Keep sawing, all the while paying careful attention to the *hinge*, the piece of uncut wood between the back cut and the undercut. The tree will topple before the saw cuts all the way through, and how it falls will depend largely on the hinge. As you saw, try your best to keep the hinge uniformly thick. If it is uneven, when the tree begins to fall it may tear easily from the thin end of the hinge while hanging back on the thick end, perhaps causing the tree to twist in its fall.

In addition to the hinge, you should also keep a very close watch on the *kerf*, the space that the saw leaves behind it as you cut through the tree. This space will give you your only advance warning of how the tree is going to fall. When you get about one-third of the way toward the undercut, you should notice the space getting ever so slightly bigger. Good! This means that the tree is beginning to pull toward the undercut, which is where you want it to go. Keep sawing until you hear the crack. Remove the saw, back off, and — as if I had to tell you — stay alert.

On rare occasions, however, you'll notice that the kerf, instead of getting bigger, is closing up on you. This means that you have misjudged the lean or the balance of the tree and that the tree has no intention whatsoever of falling in the direction of the undercut. Don't just keep sawing in the hope that the tree will change its mind; trees don't change their minds. If you keep sawing, the kerf will eventually close up so tight that it will trap the saw. Before this happens, remove the saw and put tension on the tree with a rope or a forked stick to force it over to the right direction. Or, if you have them (and you should), knock some wedges into the back cut to open up the space. One way is to saw a bit, then, keeping the saw in place, knock the wedges in, saw a bit more, then knock the wedges in a bit further — until as the tree is weakened it's lifted up toward the undercut. If you are going to use this procedure with a chain saw, make certain you use wooden or plastic wedges rather than metal wedges.

There is one more eventuality which is rare indeed but does happen once in a blue moon. Sometimes you find yourself sawing, the hinge growing thinner and thinner, but the tree gives absolutely no indication of which way it is heading. What's happening is that the tree is balanced on its hinge. It may fall one way, it may fall the other way, or — most dangerous of all — without a hinge to guide it, it may slide off its stump and kick out at you. If you ever feel the hinge getting too small for comfort, stop sawing right away and use ropes, a forked stick, or wedges to get the tree down.

Using an ax I've already told you that dropping a big tree with an ax is dangerous, difficult, and foolhardy — right? If you decide on going ahead anyway, here's how. Make certain, first of all, that you are working with the lean of the tree, that there are no dead branches to get jarred loose, that the dangers of mis-

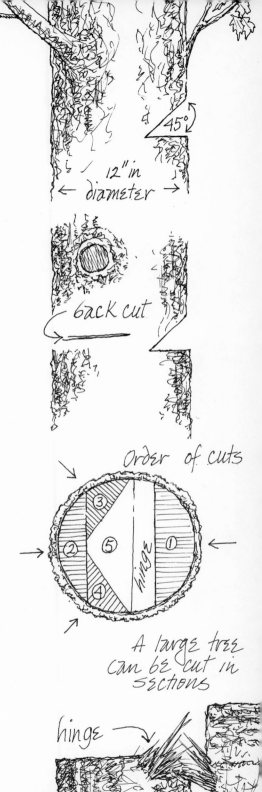

45°

12" in
← diameter →

back cut

Order of cuts

A large tree
can be cut in
sections

hinge

calculation are minimal, and that you have your escape routes well prepared.

Then begin chopping an undercut on the side where you want the tree to fall. But instead of making a cut one-third of the way through the tree, you should chop about 50 or even 60 per cent of the way through. Also, unless you're extraordinarily proficient, you won't end up with a neat wedge but with a wide, sloppy cut. This wide cut should have a *focus*, that is, one area running across the tree that is deeper than the rest. Try to leave this focus as clean and even as you can, since it will be serving as a hinge when the tree falls.

When you've completed your undercut, get around to the other side of the tree to begin your back cut, or felling cut. Aim it two or three inches above the focus of the undercut. And do your best to keep the hinge especially thick on both ends. If you give in to the temptation to chop away at the easy corners rather than the bulky middle, you will weaken the edges of the hinge in such a way that when the tree falls it might twist on its eccentric hinge.

Felling with the lean Sometimes you come upon a tree that is leaning with an exaggerated, theatrical posture off to one side. Trees like this can be easily felled with a single cut on the opposite side of the lean. When you have cut halfway through the tree, it will usually crack and fall, leaving the bottom half attached. You can sever the tree entirely, of course (watch out for the roll). Or you can let it be. The connected part will keep the leaves and twigs alive for a season or two, providing excellent wildlife browse at a level where deer and rabbits can benefit from it.

Lodged trees Sometimes you cannot get the tree quite into the bed you have made for it (a fairly common problem in other areas of life too), and the tree gets hung up in the branches of another tree. If this happens to you, you have troubles. There is only one safe way of dislodging it: wrap a cable around the butt end and use a winch, a "come-along," or a truck to pull it free.

Sometimes you can "walk" it away, using another pole as a lever. If none of these tricks work, leave it in place. Call it a natural bridge, a deer tunnel, an Arc de Triomphe, a half-finished tepee — call it whatever you want, but if you can possibly avoid it, don't try to chop it down. It will only get you into more trouble.

Stumps After you fell a tree, you are left with a waist-high stump. You might want to level the stump to the ground, for aesthetic reasons if for no other. (If you're using a chain saw, watch out for stones lodged in the base of the tree.)

Whether you level the stump or not, the main thing to beware of is the possibility of crown sprouting. Many trees will wait for a few months after you've chopped them down; then, after you've safely forgotten about them, they suddenly come out with a circle of little sprouts or suckers arranged like a game of ring-around-the-rosy around the old stump. Instead of one tree, you'll have a dozen. It's like Hydra.

Your wildlife, of course, will probably rejoice at all the tender young sprouts and buds so close to the ground. If, however, you are making a trail or creating a clearing, you may not be quite as happy. The officially recommended way of treating a stump is with any of several poisons on the market. Some of them don't seem to work very well, others seem to work too well, and all of them are dangerous and unpleasant to handle. If you don't want to use them, you'll have to resign yourself to maintenance. Return every four or five months for a few years to lop off the suckers; eventually the root system will get the message and will give up.

Safety I'm usually pretty cool and unconcerned about safety, even when I was dealing with kids. I could, I suppose, have given them a long, dull rap about safety, nagged at them, and spied on them; and if I had kept a sharp lookout, I suspect that I would have had the experience of watching them get hurt. Nagged-at kids always seem to be the ones who get hurt. On the other hand,

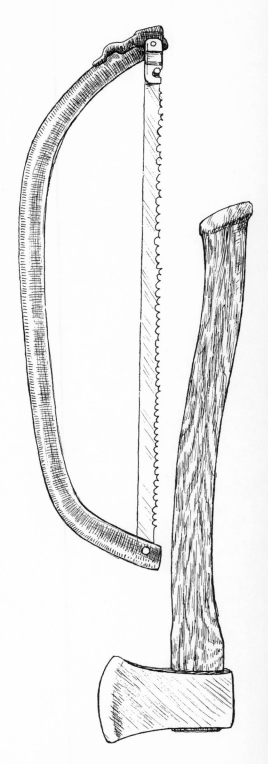

I like to feel that happy kids enjoying lots of freedom and responsibility are far less likely to have accidents.

But when it came to felling trees, my cool sort of dissolved. Sometimes I'd go out on a clear morning with a group of kids, very much aware of being warm and alive, of our friendliness, our good spirit, our feelings of adventure — conscious of ourselves as a community of humans about to have fun and serve the needs of our wild land. Then I'd look at the saws and axes in the hands of a group of fourteen-year-old kids, and I'd get scared at the thought of a serious accident.

There were seven precautions that I always took and I suggest you do the same — whether for a group of kids, friends, or for yourself.

1. Avoid windy days.

2. Stay away from steep slopes.

3. Take along a first-aid kit and make sure you have a vehicle nearby.

4. Issue hard hats and insist that they be worn. I had expected some resistance about hard hats. I thought that kids would feel toward them the way they sometimes feel toward seat belts. But I was wrong. Kids love hard hats; it's a symbol that they are doing dangerous, adult work. They not only wore them faithfully in the woods, but they wore them during lunch, paraded around in them, modeled them happily, and more than once tried to take them home at the end of the day.

5. Keep the groups small — under five kids. When you are dropping trees, you have to spread out to avoid dropping them on each other. More than five kids will be so spread out that you won't be able to keep track of them.

6. Never fell trees alone.

7. Learn to stop early. Felling trees is hard work. Whether you are using saws or axes, you soon develop sore muscles and blisters. After a while — sometimes after a very short while — you find yourself holding your body in weird ways to avoid

straining already tired muscles. Soon you begin holding your ax or your saw gingerly to avoid breaking blisters. When you notice this, stop the logging right then.

These are my seven safety rules, and I followed them scrupulously. They worked for me, yet I was still never very easy about felling trees with kids. On one hand I wanted to give them a maximum of freedom — that's what wild lands mean to me. I wanted them to have fun, take chances, and attack problems with imagination, with spirit, and without any adult hovering over them anxiously. On the other hand I wanted them to be safe and do things "properly." This put me in a double bind. Usually, I sinned in the direction of freedom. The kids made out very well this way — no one ever got hurt and the kids were quite happy — but at the end of the day I was usually a total wreck.

READING

The U.S. Forest Service has a variety of booklets about logging, many of them aimed at the farmer with a small wood lot and not too technical. There are many of them, for the different areas around the country. The best of those I've seen is listed below.

Northeastern Loggers' Handbook, by Fred C. Simmons. Handbook No. 6. Washington, D.C.: U.S. Department of Agriculture, 1951.

This is a 160-page book, clearly illustrated and written for "the young or inexperienced woodman." Since it is aimed at the logging industry, you'll find sections about logging trucks, road construction, cable logging, etc., that are not very useful but are interesting nevertheless.

There are very good sections on the use and care of axes, saws, and other tools; on how to fell trees; on overcoming unfavorable leans; and (if you're interested) on limbing and bucking the tree into usable logs.

It's well illustrated and simply, plainly written.

Axe Manual of Peter McLaren, by Peter McLaren. Philadelphia: Fayette R. Plumb, 1929.

This is a real oldie, but if you happen to spot it in some fusty old bookstore, grab it. It dates back to the pre-chain-saw days when the ax was the woodsman's bread and butter. McLaren was "America's Champion Chopper," and the book gets right down to the nitty-gritty of how to hold the ax, how to make the chips fly easily, the different kinds of cuts you can make, wedges you can improvise out of log sections, and lots of other very useful information for people who still use an ax. There is also a good section on the care and sharpening of axes and on how to refit handles. It was published by the makers of Plumb axes.

ᴄᴧ3 *The Mulch Mystique*

Stripped of its mystery, mulching is a very simple act. You locate a bunch of organic matter, like straw, leaves, lawn clippings, or wood chips; you collect it from where it's not wanted; and you spread it out someplace else where it will do some good.

Mulching is a very simple project — dull and repetitive if you take it too seriously or keep at it too long, but lots of fun for yourself or for groups of little kids who can roll in the mulch, horseplay in it, and throw it at each other while (incidentally, it seems) doing a piece of important work. Not only can it be fun, but if you think deeply about mulch, it can be a perceptually liberating experience as well.

These may sound like exaggerated claims, but there is something about mulching that generates fanatic devotion. Enthusiastic mulchers — and there is no other kind — insist that with the magic of mulching their gardens no longer need weeding, watering, or fertilizing as ordinary gardens do. Yet they grow king-sized cauliflowers, luxuriant lettuces, peas that ripen weeks before their neighbors', and squashes that stay firm long past the first frost. When I first heard about mulching, it sounded very interesting, very seductive, and very far-fetched. I didn't believe it!

In fact, for many years I had the vague idea that mulching was an exotic invention of organic gardeners. Or, to be more precise, the invention of a single organic gardener, an eccentric New York escapee named Ruth Stout who settled in Connecticut, wrote a book and several articles about mulching, and has become, so to speak, the *grande dame* of mulch.

But recently I have learned a lot more about mulching — especially about mulching on wild lands — and I've discovered that mulching was invented long before Ruth Stout. In fact, it was invented long before the human race. Mulching is a marvelous process invented by plants — a process by which wild plants have survived and prospered for eons. Without watering, without weeding, and without fertilizing.

The mulching process is actually very easy to understand. Every kid who has walked through the autumn woods, swishing and kicking the piles of dry leaves, knows something about mulch. Someday, attack one of those big piles of leaves, and as you scrape away at the various layers, your eyes, fingers, and nose will tell you everything you have to know about mulch.

The topmost layer consists of dry, fresh-fallen leaves — whether the broad leaves of deciduous trees that fall all at one time or thin pine needles that drop throughout the year. Brush these dry leaves aside and beneath them you will find the older leaves — matted, moist, and half decomposed. Like an archeologist, keep digging down through time. The leaves will get more and more decayed until finally you reach the *leaf soil*, or *humus* — the thoroughly rotted leaves and twigs that earthworms and moles have already begun to mix with the mineral soil underneath.

There are more lessons to be learned from leafing through good soil than leafing through a good book. You can see how the area beneath a well-mulched tree lacks weeds and competing grasses, their seeds buried under tons of leaf litter. Dig your fingers under the humus and you can feel how the thick layer of leaves acts as a blanket, keeping the earth from freezing too deeply in the winter and from baking in the summer. Feel how squishy and spongelike the leaf mold is — a perfect texture for absorbing and holding water while allowing air to circulate. And you can easily understand how the leaf fall is actually a recycling of nutrients: the natural nitrates and other more exotic minerals that the roots have mined from the lower parts of the soil go into the leaves, and at the end of the growing season, they are returned to the soil for use again in the following years.

These are very important concepts, and they help one to understand that mulching is a natural process rather than some magic, hocus-pocus piece of untidy mystification. But as you dig down into the leaf mold, I think you will discover something more amazing than just the physical properties of the mulch. This world of decaying leaves is, in itself, a marvelous thriving environment of incredible complexity and beauty. If you are lucky, you may stumble upon the giants of this environment: the salamanders, mice, and moles that tunnel through the humus. You are more likely to find the larger bugs, such as centipedes and pill bugs. But look closely, with a lens if possible, and you will *definitely* see a thriving, teeming world of creeping, crawling, humping, wiggling, thrashing, squirming beasties — thousands of them to the square foot. Some are barely visible motes gliding over a leaf edge; others are incredibly fierce and dramatic, like miniature dragons in a Chinese parade. This is a rich, full world — yet it is only a hint of the fantastic microscopic universe hidden to our eyes.

I feel that there is as much wonder, beauty, and mystery in this terribly alive environment as there is in the unexplored jungles of the Amazon. When I am with kids, I try to communicate this wonder to them, to make them realize that when we mulch a tree, we are not merely doing some mundane agricultural act that will benefit the tree and improve the soil. We are creating a natural environment that is very wonderful, very complicated, and forever beyond our understanding.

If you are dealing with kids, I urge you to get them to help you dig around in the humus. Most kids love to dig in dirt, and this is the most interesting dirt in the world — dirt crawling with decay, earthworms, and bugs. But some kids have been brainwashed by the Mr. Clean sterilize-the-world fanatics until they think that dirt and bugs are bad. This is tragic. Do your best to help them overcome it. If the kids are too uptight to dig around with you, try anyway. The knowledge that they have met some adult, however crazy, who likes dirt may someday help them in their quests to become healthy, accepting human animals.

To mulch or not to mulch? Before you run off madly from farm to farm begging for spoiled hay, rotten manure, and other such goodies, take a long, slow look at your forest floor. You can learn more about the health of your forest from the depth of the humus at your feet than from the height of the trees above your head.

If the layer of leaf mold and humus is thick, rejoice! Your forest is in natural, healthy condition and you do not need mulch — in fact, if your forest is in that good condition, you do not need this book.

On the other hand, if you are restoring a piece of abused land, you are probably not so fortunate. Fire may have burned the leaf mold away; people, vehicles, or cattle may have pulverized or compacted it; or you may have just planted a new forest where the trees have not yet begun to build up their mulch beds.

Any bare — or balding — piece of land can benefit from mulching. Placing mulch around a tree is both helpful and natural. Other places that can use mulch, urgently at that, are bare slopes or disused road beds (see Erosion Control, pages 75-76) and around trees suffering from compacted soil (see pages 57-58).

What kind of mulch? The longer you mulch, the more prejudices and stubborn attitudes you will probably develop, until you eventually become so crotchety, opinionated, and impossible that everyone will consider you an expert.

My own hang-up was rotten horse manure. I attributed all sorts of magical properties to The Stuff. I knew of several stockpiles from old stables, and I cherished these bits of knowledge as a prospector might cherish his mines of gold.

Yet, to be honest about it, any vegetable matter will do as a mulch. Some, of course, are better than others. Some are more acid, others more base. Some have more nitrogen than others. Some rot faster and others more slowly. If you are raising vegetables, delicate flowers, or exotic shrubs, you might worry about these fine points. But for general wild-land management, any organic matter is better than bare ground. What you use will probably depend on what you can get. One word of warning, though: do not use an inflammable mulch in a high-fire-hazard area. Dry sawdust in particular has been known to ignite by spontaneous combustion.

Here is a list of some of the more commonly available mulches, and a few comments.

HAY Look for "spoiled hay," that is, hay that has been ruined by rain and is no longer good for animal feed. The farmer's woe can be your delight. You can often get spoiled hay free for the hauling. It makes a fast-rotting, excellent mulch. It is easy and pleasant to handle, and it is the best mulch you can get for erosion control (see pages 75-76).

LEAVES Obviously the most natural mulch for trees. Before collecting them from city streets or parks, think about how to compress them in the back of your truck. Some sheets of plywood and a few stones might help.

Oak leaves and pine needles have a reputation for making the soil acid. I generally wouldn't be too concerned about this, but if you mulch often with these leaves, an application of lime won't hurt any.

LAWN CLIPPINGS This is perfectly good stuff, and city parks departments will often give it to you free. Don't lay it on much thicker than six inches, because bigger piles of green lawn clippings sometimes heat up ferociously as they decay.

SAWDUST   Personally, I hate sawdust. I find it hard to handle and boring to shovel, it looks dull and sodden when wet, it tends to blow away when dry, and it makes me sneeze. Other people use it regularly, and some people even prefer it. Different strokes for different folks, as they say. If you live near a sawmill, you may get it free and help reduce air pollution at the same time.

One occasional problem with sawdust is that sometimes the first application creates a nitrogen deficiency in the soil. It seems that the organisms that digest cellulose use up lots of nitrogen, which they borrow from the soil. Once the sawdust starts to rot, however, the organisms die off and return the nitrogen to the soil, so the problem is short-term. If a recently mulched tree shows signs of nitrogen starvation (look for a yellowing of the leaves), add some sort of nitrogen-rich compost or fertilizer.

WOOD CHIPS   Wood chips are another of my favorites. They don't decay as fast as sawdust and have none of sawdust's drawbacks. They are often available from utility companies or anyone else with a chipper.

OLD CHRISTMAS TREES   Why not? Returning a Christmas tree to the soil would be an excellent Christmas present for your land. And soliciting Christmas trees from the community is a fine way of getting people involved.

DRY WEED STALKS AND HAY   If you are planting trees in the middle of an old pasture, you might bring along a sickle or scythe and cut weeds and hay right from the site. Don't worry about weed seeds. They won't stand a chance if you make the mulch deep enough.

AGRICULTURAL AND FOOD PROCESSING WASTES   Depending upon your local agriculture and industries, you might get buckwheat hulls, ground corncobs, rice or wheat stalks, crushed sugar cane, peanut hulls, spent brewery hops, cocoa bean hulls, or well-rotted manure. Look around, become mulch conscious, and remember: *any* organic material (no matter how weird it smells) is better than bare soil.

Applying the mulch: There's really not much to it. If the soil is hard and compacted, you might rototill the area first. Otherwise, just dump the stuff around the tree. Don't even dig up the weeds and grasses; unless the sod is exceptionally thick, bury them! If the mulch is deep enough, the weeds and grasses below will rot and add to the mulch.

After you've dumped the mulch and spread it around a bit, you must rake it away from the tree trunk. If you leave it piled around the trunk, the millions of little beasties will soon arrive to nibble away at the bark. Spread the mulch so that most of it is under the *drip line* — the area underneath the outermost branches of the tree. If the material is light, you can throw some sticks, branches, or logs over it or wet it down so that it won't blow away.

For erosion-control mulching and mulching on steep slopes, see pages 75-76.

Afterwards After you've mulched, there is nothing else to do. Nothing. Please don't cultivate the ground with the idea of working some of the mulch into the soil. If the mulch has not yet decomposed, cultivating won't do any good and might even steal moisture and nutrients away from the roots. If the mulch has decomposed, then it is soil, organic soil, and the earthworms will have an absolutely ecstatic time mixing it with the mineral soil underneath. If you want your share of ecstasy, there is nothing better you can do than watch the earthworms.

Kids As far as the land is concerned, mulching is always a very successful project. But I've found that many city kids who come to the park are severely puzzled by it all. Nature for them is often a still life. Their idea of a beautiful forest is some sort of Rousseau-ian painting of eternally perfect trees and clean green lawns upon which romp Bambi-like animals who never defecate or die. Their idea of a worthwhile conservation project is to pick litter. But now, instead of making the woods clean-clean-clean, pretty-pretty-pretty, they find themselves following a madman through the forest, spreading garbage under the trees, and making homes for millions of yucky, ugly wiggling things. If this wasn't shocking enough, the magic mulch they are being asked to handle turns out to be old horse manure!

I always try to treat kids with respect. Their fears and prejudices are very real to them and very important. I've learned that no matter how clear my own vision may be, I cannot force them to *see* when seeing is too threatening to their whole way of life.

If you find yourself with a group of kids who are too uptight to handle mulching, try to explain what it is all about as clearly and gently as possible. Tell them what you feel, let them absorb intellectually what they can, and then go on to something else. No matter how sure you are of your own sensibilities, you cannot force them on anyone else — especially kids.

As you have probably guessed, I've had some difficulties with mulching projects. But I've also had a lot of fun. Many kids have really gotten behind throwing the mulch around. One common reaction has been to treat the whole business of spreading manure and making earthworm houses as something hilariously funny. I've seen kids giggle for hours over it. In fact, they have occasionally made ruthless fun of me for liking such things, but they were so obviously turned on and were so obviously thinking, feeling, and experiencing that being made fun of seemed more like a reward than an insult.

READING

I haven't found any books specifically devoted to mulching on wild lands. The following books deal with garden mulching, and they will certainly give you many good ideas.

How to Have a Green Thumb Without an Aching Back, by Ruth Stout. New York: Cornerstone Library, 1955.

This is a classic of a book, available in paperback and always in print. I expect it is one of the few modern books that will be in print 500 years from now. It's thoroughly wise, eccentric, individual, and very funny. It makes wonderful reading, even for nongardeners — just as Izaak Walton's *The Compleat Angler* is a delight to those who never fish.

Handbook on Mulches. Handbook No. 23. Third printing. New York: Brooklyn Botanic Garden, 1970.

Like other handbooks in this series, this is a hodgepodge of articles by different authors. Some are very helpful, some really excellent, and some rather esoteric. There's a very good section on sawdust mulching. It can be ordered by sending $1.50 to Brooklyn Botanic Garden, 1000 Washington Avenue, Brooklyn, N.Y. 11225.

Soil Animals, by Friedrich Schaller. Ann Arbor: University of Michigan Press, Ann Arbor Science Library, 1968.

This readily available paperback was written for the "layman" by a German zoologist. It deals mostly with microscopic animals and will turn you on to an utterly fascinating world. There's a good section on how soil animals turn leaves into humus, with lots of information about earthworms and all sorts of strange facts about the way the animals breathe, eat, feel, smell, and mate. It reads like a combination of a college biology text and *Ripley's Believe It or Not*.

✎4 *Be Your Own Tree Doctor*

This chapter describes several different things you can do to cure a sick tree or heal an injured one. But before you do anything, I hope you'll stop to consider that on wild land, diseases and wounds play valuable roles. There is nothing *bad* about a diseased tree. In fact, sick, dying, and dead trees are necessary to a balanced environment.

Let's say, for example, that you have an oak tree that's infested with oak moths. It is obviously a sick tree, and, according to professional foresters and landscapers, it must be sprayed at once. But if you look at it without prejudice, if you really make an effort to see the tree as it is, you will see much more than just a sick tree. You will see a healthy, thriving, vibrant colony of oak moths — wiggling, spinning, and fluttering, totally admirable and unspeakably, joyously alive.

Oak trees can generally withstand two or more years of severe moth attacks without being seriously affected. But let's imagine the very worst — namely, that the tree will die. If this is the only tree shading your back yard, you might very well be alarmed at the prospect. But in the wilds, is the death of a tree really so terrible? Dead trees house woodpeckers, small birds, owls, squirrels, raccoons, and possums. Tear away the loose bark from a dead tree, and out crawl termites, beetles, and flies from among colorful explosions of slime mold and other fungi. Once I found a sleepy, confused bat underneath a flap of bark. Indeed, sometimes it seems that a "dead" tree is more gloriously alive than a healthy tree. The death of a tree means that tree matter gets transformed into termite matter. In a balanced environment, death is not The End but an ongoing process — and a very lively process at that.

In working with kids, I found that they have very powerful feelings about disease, injuries, and death. They'll scarcely notice a forest full of twittering, fluttering birds; but they'll never fail to gather around the body of a dead finch or injured pigeon to stare, commiserate, and wonder. Whether it's a bird or a tree, kids seem to have an open and admirable curiosity about something dead or dying. They have a wonderment about death which I think is basically a wonderment about life. It was a pleasure not to fill their heads with sanctimonious, anxious thoughts, but instead to share in their wonder.

The death of trees in the wilds is to be expected, and it is not especially to be mourned. A dead tree is not a tragedy, and a sick tree is not an emergency that calls for sprays and poisons. If a tree is naturally troubled, it would be condescending to meddle in its fate. We should beware of foisting our own images of health and prosperity onto the environment.

Yet there are times when caring for trees is necessary — especially when you are undoing the damage done by people. I was constantly treating injuries caused by jackknives, automobile bumpers, bulldozer blades, snowplows, bad pruning (especially by utility companies and road crews), soil compaction, barbed wire, or cows with itchy heads. In short, there is plenty of opportunity for tree doctoring on wild lands — as long as it's in the spirit of *undoing* rather than *doing*.

TREATING AN INJURED TREE

This section deals with how to treat wounds, gouges, tears, and other physical injuries to a tree. Treating an injured tree will help it considerably, but there is rarely anything urgent about it. Most trees are remarkably hardy and remarkably slow. They are slow to mature and slow to die. We have all known trees that were rotten through and through, riddled with holes and crawling with bugs. Every fall the tree would be given up for dead, and every spring it would delight us with a burst of brave new leaves. Most trees are very persistent, so relax. Wait until you're in the mood for a leisurely, intimate, craftsmanlike communion with a tree. The tree will wait, and so can you.

Why treat a wound? When you come across a freshly wounded tree, you may find sap and resin oozing out and you will think, Good grief, the poor tree is bleeding to death! Don't worry. Trees rarely bleed to death — otherwise the maple sugar industry would have dried up long ago.

The problem with wounds generally is that they are places where moisture can gather. And rot sets in wherever there's moisture. So the principle behind treating tree injuries is to eliminate any place where water might collect.

First aid If you catch the injury fairly soon, before rot sets in, you can simply give the tree a little first aid.

If the injury is in the form of a jagged stub, cut the stub off, following the instructions for pruning on page 122.

If the injury is a wound directly on the trunk or against a major branch, use your jackknife to trace an elongated, pointed ellipse — like a football standing on end — around the whole wound. The knife should be cutting through firm bark all along its path.

Next clean the area within the ellipse. Remove all the bark, fibers, and shreds. You can use a broad stroke of the jackknife, a broad chisel, a paint scraper, or a file. This will most certainly cause suffering to the tree, so make sure you tell the tree it's for its own good. Trees can be extremely aware. Watch your bedside manner.

Whatever tool you use, get the area within the ellipse down to smooth bare wood. Don't leave islands of bark within the ellipse, no matter how sound the bark looks. Remember: smooth! Pretend you are a drop of water sliding down the trunk or splashing directly onto the

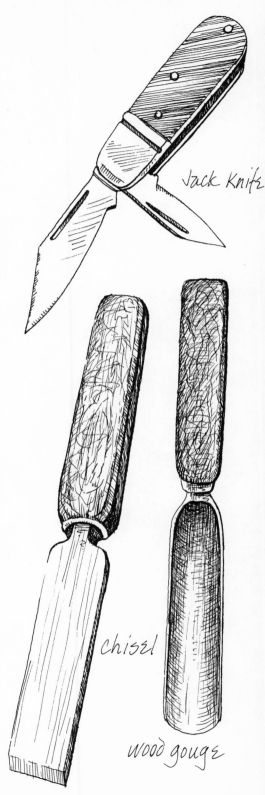

Jack Knife

chisel

wood gouge

wound. Gravity is pulling you down. Is there any place you can stop and rest? If so, that place must be removed.

Next use your knife to bevel the edges of the ellipse outward. Look at what you have done. Between the bark and the exposed wood you'll notice a thin, juicy layer. This is the cambium, the tree's point of greatest aliveness. This is the part of the tree that grows and will form the *callus* or scar tissue that, over the next several years, will creep over the wound, covering it and protecting the tree from rot.

You can then apply a dressing if you want to. (It's not essential.) Cover the edge of the ellipse with orange shellac to prevent the cambium from drying out. Shellac is a tree's very own dressing, made by pine trees specifically to seal their own wounds. Then swab on the tree paint — unpleasant, asphalty stuff that will help seal out moisture. It's available in most garden shops.

After you have applied the dressing, look over your work. You have turned an ugly wound into a neat scar. Congratulate yourself on a fine, craftsmanlike job. Over the years, if you watch for it, you will notice the callus forming over the scar, closing it off until the wound has completely disappeared.

Surgery If the wound has been neglected for a long time and rot has set in, it's too late for first aid. You'll now have to perform a bit of tree surgery.

By tree surgery I do not mean that you should fill cavities with cement. This business of filling cavities is not tree surgery — it's more like tree dentistry. Frankly, I join with the wrens, the wood ducks, the squirrels, and the bees in rejoicing every time I see a hole in a tree. In fact, many tree surgeons have concluded that cement fillings are often harmful. Cement is rigid, while a tree is always bending and swaying. No matter how carefully the cement is put in, it will often crack, and cracks are ideal places for rot to get started.

To treat a rot-infested wound, you must first get rid of all the punky, rotten wood. But before doing anything, probe. With a brace and bit, drill a few test holes into the rotten wood and keep drilling until you hit solid wood. Be especially sure to probe downwards, which is how rot usually spreads. Does the rot extend deep into the tree? If so, wish the tree good luck, pack up your tools, and go on to something else. You can literally spend days cleaning out a large cavity and never get to the bottom of it.

If, however, you hit solid wood within six inches or a foot of the entrance, you can begin to clean out the rotten wood. The best tool for the job is a woodpecker's bill — attached to a woodpecker, of course. Otherwise, use a wood-carving gouge. If the wood is really rotten, you can just shovel it out, using the gouge as a spoon. Don't feel you're hurting the tree: rotten wood is no more alive than are dried, fallen leaves. Dig in as far as you can, using the palm of your hand as a hammer against the gouge. Work the hole down to solid wood. Then smooth the inside as best you can, again eliminating shreds and pockets where water can collect. Without the special equipment that a tree surgeon has, you'll never get the inside of the cavity perfectly smooth. So leave perfection to the gods and the masochists. Just do the best you can.

After you have gotten rid of the rot, you are left with a cavity. Next you *must* provide drainage. Otherwise, the hole will collect water and become a small pond — picturesque, but "rot city" nevertheless. If the hole is a small one, you can simply slope it outwards so that the water drains through the mouth of the hole. If the cavity is deep, however, you will have to drill a hole from the outside of the tree, angling it up toward the bottom of the cavity. Make sure the drainage hole hits the cavity at its lowest point. It is considered good practice to knock a length of pipe or copper tubing into the hole to prevent the tree from sealing it up within a few years.

Finally, you can apply the dressing. Professionals disinfect the cavity with bichloride of mercury, but I find the stuff much too dangerous for my own comfort. Instead, I use denatured alcohol, which seems to do a fairly good job. (Also, at the end of the day I clean my tools with the alcohol so that I will not carry an infection to the next tree I work on.) If the tree is particularly valuable to you, you might then smear a layer of tree paint throughout the cavity, being careful not to plug up the drainage hole. But as long as the sides of the cavity are smooth and the hole is well drained, I don't think that tree paint is necessary.

Cleaning out a rot-infested wound has a double benefit. It stops the spread of rot, thus helping to keep the tree sound. It also creates an airy, well-drained cavity — a Waldorf-Astoria of a cavity — for which some lucky animal will be profoundly grateful.

BARBED-WIRE WOUNDS

When I was working with kids, we tore down a lot of barbed-wire fences. This was an especially good project for groups of older, tougher kids. They felt they were doing something useful, something easy to understand, and (best of all) something destructive. In case you've forgotten, kids just love to destroy. And what better thing is there to destroy than a barbed-wire fence? We all hate fences, all fences, but especially barbed-wire fences. Between us we have all ripped thousand of pairs of pants and gotten millions of nasty scratches from sneaking under or over barbed wire. I think of this project as The Ripped Shirt's Revenge.

That we hate barbed wire and that it looks so ugly are reasons enough to remove it. But there's more. By tearing it down, you will be saving the lives of hawks and owls, who often get tangled up and maimed while hot after a rabbit or a mouse. Finally, all along the fence line you will most probably find (and save) trees suffering from barbed-wire strangulation because they were used as fence posts.

There is, however, one circumstance in which I temper my hatred of barbed-wire fences. Sometimes, by protecting the ground beneath it from grazing animals and by providing a perch for many birds who drop seeds, a barbed-wire fence transforms itself into a hedgerow of interesting vegetation. This creates a valuable wildlife habitat, especially in otherwise open meadows. In such a case, I would leave

the fence in place, although I'd still walk its length to check for injured trees.

Removing the fence   There is only one way to handle barbed wire — carefully and with gloves. And make sure you have thick gloves. The scratches you get from rusty barbed wire have a way of staying around for a long time, itching, threatening to get infected, and sometimes carrying out their threat.

First, detach the wire from the posts — usually with a pair of pliers. As you detach the wire, begin rolling it around something, a stick or a piece of the fence post, perhaps. If, instead of rolling it, you just bunch it, the consequences will be dire. After a few hundred feet, you will find yourself with a tangled heap of barbed wire that will dwarf you, your truck, and your powers of invention. You can't leave it behind because it's too ugly; you can't drag it over the ground because it keeps getting caught up in everything; you can't load it into the truck because it's too springy; and you certainly cannot untangle it in anything less than a month. Hassling with balls of snarled barbed wire is a most discouraging (and painful) way to spend an afternoon. So keep rolling it. When the spool gets too big, cut the wire and begin another spool.

After taking care of the wire, you can remove the fence posts. Dig them out using bars, sledgehammers, and shovels. Used fence posts are salable or tradable if they're made of metal or some weathered, durable wood.

Finally, after you have done battle with the wires and the posts, you can return like a medic to treat the wounded trees.

Unstrangling your trees   Barbed wire wrapped around a tree is something like a hanging, except that the noose does not tighten. Instead, the noose stays fixed, while over the years the tree expands until circulation gets cut off. The tree almost invariably dies. This is a sad and gruesome death — all because someone was too lazy to knock in an extra fence post.

If the wire is recent, you can probably pull it off easily with pliers. If the wire is embedded in the bark, however, stuck in by resins and sap, you will have to be more careful. Grab one end firmly with the pliers and pull steadily away. Don't wiggle or make a sharp bend in the wire; otherwise the wire may break off. Just pull steadily and hard at a shallow angle until something happens. If you're having a good day, that something will be the release of the wire.

Let's say, however, that you're not having such a good day and the wire breaks off so that you no longer have a place to grab it. If the wire was merely stapled to one side of the tree (rather than wrapped completely around it), I would leave it in place. The tree will suffer from this piece of wire, but it will suffer even more from your gouging to get the wire out.

But if the wire was wrapped entirely around the tree, you will have to remove it, or at least sections of it, to save the tree's life. Dig in at several places with a chisel to expose parts of the wire. Break the wire with a sharp blow of the chisel, grab one strand at a time with the pliers, and yank out what you can from each side of the gouge. It is not necessary to get all the wire out. Just get out enough to prevent complete girdling of the tree and to keep some circulation going. After you're through, treat your gouges as wounds, apply a coat of tree paint (not essential but helpful), pat the tree on the trunk, apologize for the human race, and hope for the best.

Afterwards   There is a joyful surge of energy that comes after a day of ripping down barbed-wire fences. You have taken a major step in uncivilizing a piece of land. You have liberated it, removed the signs of its bondage, and begun to heal its scars. You will soon begin to see your land in a brand new way, without its fragmentation into pastures and old property lines.

At the beginning of the day you were fresh and clean, while the land was crisscrossed with barbed wire. By afternoon the land will look almost reborn — but you may very well be crisscrossed by nicks and scratches. Barbed wire dies, but it dies fighting. Please treat yourself (and whoever happens to be helping you) at least as well as you have treated your trees: use soap, hot water, and proper first aid for everyone's cuts.

BULLDOZER DIRT PILES

Here is a project where you tag along after a bulldozer, trying to undo some of its damage.

In the tree-planting chapter that comes later, I emphasize how necessary it is to transplant a tree with its crown (or root collar) at ground level. If the tree is planted too deep in the ground, its roots suffocate and its trunk rots.

Bulldozer operators on the whole don't seem to understand this. Wherever they have been doing road work, trail work, or grading, you can be sure to find dirt piled around the bases of trees — just as if this were the very best possible place for it.

When you come upon trees drowning in dirt, pull out your shovel and start digging. Keep at it until the area around the tree is back to ground level. You should do this for any tree, but it is especially important for conifers and for white oaks, tulip trees, lindens, and beeches, all of which suffer easily from trunk rot.

COMPACTED SOIL

Soil compaction around trees is a very serious and persistent problem. It happens when the soil is compressed by bulldozer treads, vehicles, foot traffic, cattle hoofs, or careless agricultural practices.

Whatever the reason, trees suffer because air cannot penetrate through the compressed soil and water cannot seep through. Or if water does seep through, it doesn't drain properly.

There's no easy cure for compacted soil, but there is a lot you can do to help get the healing process started.

How to recognize compacted soil If you're not sure whether or not your soil is compacted, try this test. Dig a hole two or three feet deep. Then fill it with water. How long does it take the water to drain? If it takes a long time — say, twelve or more hours — your soil is either compacted or fantastically clayey. In either case, the treatment is the same.

Treatment First break up the upper parts of the soil. Cultivate it as deeply as you can. Hoes, picks, mattocks, and muscle power will do the job, but a rototiller or tractor (if you have one) will do it with less complaining.

Then lay a thick bed of mulch over the soil. Lay it on tenderly, like a blanket. Don't work it into the soil. Just let it sit and rot. (See Mulch Mystique, page 46.)

The best long-range cures for compaction are earthworms, gophers, moles, ground squirrels — and time. Time will pass at its own rate, without any help from you. You can plant earthworms under the mulch, but if you don't, they'll probably arrive anyway. I don't know how or from where, but they'll arrive. Moles, gophers, and ground squirrels are marvelously independent. People who hate them can't seem to get rid of them, and people who want them can't seem to attract them. But maybe you'll be lucky.

In any case, all you can really do is keep mulching and wait. The mulch will rot from above, the tree's roots will break up the earth below, and the earthworms will happily mix it all together. It will be geological ages before the soil gets back to "virgin condition" of good tilth and high organic content — but once you break up the compaction, cover it with mulch, and keep away whatever caused the compaction in the first place, you have begun a soil-building process that will continue by itself until the next ice age.

FEEDING A TREE

If the soil is compacted, if the tree has been surrounded with asphalt, or if the leaves have been raked away every year, the tree may be a victim of starvation.

The only permanent cure for starvation is mulch (see pages 39-48). But mulch (like other natural processes) is often slow. If you are restoring a piece of land that has been badly abused and the trees are in trouble, you might want to give them an emergency shot of fertilizer to keep them going until the mulch begins to do its thing.

How to identify a starving tree   If the tree is in obvious difficulty, riddled with disease, crawling with insects, sporting dead branches and "stag horns," suffering from *tip dieback*, and looking generally beaten, forget it. It has already dug too deeply into its reserve strength. Try to think of it as picturesque, if you can, and let it die in peace. Plant some other trees around it to take its place.

If the tree is merely nonvigorous — listless, sort of droopy, and depressed looking — it can benefit from a dose of fertilizer. The signs of a sick tree are subtle. If you're in doubt, compare it with a tree that you think is healthy. Look for a slight yellowing of the leaves, smaller annual growth, and a very slow rate of callusing over wounds. Or if you're the sort of person who is attuned to "vibes," you can ask the tree how it feels and wait for the answer. But you'll have to wait around for a long time. Remember, trees are slow. So be alert, be patient, and come back as often as you can. Keep asking the same question, and if you really want the answer bad enough, the tree will find a way of telling you.

What kind of fertilizer   Chemical fertilizers are fast acting and are probably the best thing for a starving tree. If you are an organic gardener, please don't sulk. I use chemical fertilizers as medication, not food. I would not give a tree a steady diet of chemical fertilizer any more than I'd give a person a steady diet of penicillin. Mulch takes a long time to add its nutrients to the soil, however, and until it does I think a starving tree will appreciate a shot of fertilizer to keep it going.

Any general-purpose garden fertilizer will do the job. But if the leaves are yellowed, you might want to use a high-nitrogen fertilizer — say, 10-8-8 or 10-6-4 or something along those lines.

For a slower-acting but safer fertilizer, you can use more or less equal parts of phosphate rock, granite dust, and cottonseed meal. Mix these together with a large amount of compost (necessary to help break down the rock powders) and dig the mixture into the ground, approximating as best you can the following instruction.

How to apply the fertilizer Don't just broadcast the fertilizer over the ground. If you do, you'll encourage a thick, lush growth of weeds and grasses under the tree — a marvelous green carpet for those who like to lie under a tree watching the clouds, but not much use to the tree.

There is a special technique for fertilizing trees. Use bars, crowbars, or especially long bolts sharpened at one end to punch holes underneath the outermost branches of the tree (the drip line), which is where most of the feeding roots are. The holes should be about two feet deep and about two feet apart.

Weigh out the proper amount of fertilizer and pile it onto a newspaper. Use three pounds of chemical fertilizer for every inch of trunk diameter on trees six inches or wider. If the tree is between three inches and five inches, use only two pounds per inch. If in doubt, use too little. After you've weighed out the fertilizer, distribute it evenly among the holes you've made. Do not fill the holes much more than halfway. If you have too much fertilizer, dig more holes. The more holes you dig, the better off you'll be, as long as they're deep enough and within a couple of feet of the drip line.

Finally, fill the holes with dirt, stomp the dirt down hard, and water the ground thoroughly. If you cannot get water to where you're working, save the fertilizer for just before a good, soaking rain.

Many people consider early spring to be the best time to apply fertilizer. And while I personally remain skeptical, many nursery experts I know refuse to fertilize in the summer because they do not want to force new growth just before the autumn frosts.

Afterwards After you have treated a sick tree, do not expect the tree to thank you with any dramatic gestures of gratitude. It won't turn dark green overnight, grow ten feet in a single year, or do anything else exhibitionistic. Don't take it personally. Trees are very slow. They get sick so slowly that you scarcely notice it. A sick tree can live on stored nourishment for years before visibly suffering. When a tree starts to get better, the process is equally slow.

A tree does not put all its energy into new growth; it leaves such spendthrift behavior to the weeds. Instead, trees store nutrients, create a reserve, and gradually build themselves up again. No wonder they live so long.

Treating a sick tree is hard work, and as with other strenuous projects, I suggest making it short — especially if you're working with kids. I cannot see curing a sick tree at the expense of turning off a group of healthy kids. As a treat I used to bring along a few lengths of heavy rope which we'd attach to the tree's branches as a rope swing. Some people might consider this sacrilegious, but to me there is no more joyful and fitting celebration of a healthy tree and a group of healthy kids than to swing through the air on a really first-class rope swing.

Reading

Here are some helpful books about tree care.

The Care and Feeding of Trees, by Richard C. Murphy and William E. Meyer. New York: Crown Publishers, 1969.

A very helpful book, but it is definitely oriented toward ornamental trees rather than forest trees. Nevertheless, it does give lots of information, especially on fertilizing, pruning, and treating injuries. It also has good drawings.

Tree Maintenance, by P. P. Pirone. New York: Oxford University Press, 1959.

This book has gone through a number of printings and editions, and is *the* textbook and reference work on tree care. It is dull but ever so thorough. It gives descriptions of many individual trees, covering their preferences and idiosyncracies. There are very good sections on planting, fertilizing, and treating wounds. The book describes mostly "street trees," but it is still valuable for forest use. It's available in most larger libraries.

The Care and Repair of Ornamental Trees, by A. D. Le Sueur. Revised edition. London: Country Life, 1949.

This book is very British, very "ornamental," and very unavailable. It's also outdated, with a section, for example, on how to treat tree injuries caused by the German bombings. Yet it is a thoroughly knowledgeable and intelligent book — written with great care — and one of the few tree books that I find myself browsing through just for the fun of it. I don't suggest you make a special search for it, but if you happen to stumble across a copy somewhere, grab it!

❦5 *Erosion Control*

A visitor from outer space might have a good laugh at how we handle — or don't handle — erosion. Our homes have locks on the door, latches on the window, insurance policies in the dresser drawer, and we support a huge police and prison system — largely to protect a few cameras, watches, and other gewgaws. Meanwhile, outside our windows, every rainstorm carries away thousands of tons of valuable topsoil upon which we depend for our very survival. Our scale of values is pathetically confused, when you stop to think about it. With modern assembly-line methods, we could replace a stolen stereo in a few hours. Yet it takes nature almost a thousand years to rebuild one inch of topsoil.

Some people, especially farmers, have a fatalistic attitude toward erosion. Land erodes, they feel, just as people grow old, automobiles sputter and stall, and apple trees eventually give out. But land is not like that. It does not have to erode. In fact, a healthy land adds humus and builds up its fertility every year. Individual plants and animals die, giving up their lives to help build a healthy, vital, growing soil for future generations of plants and animals. This nourishing of the soil is what makes death meaningful and even beautiful. Think about that for a moment, and don't accept erosion as a "fact of life."

Another conceptual trap you can fall into is the "Grand Canyon argument." Erosion built the Grand Canyon, so the argument goes, implying that erosion is a natural process that should not be interfered with. But erosion is "natural" only in desertlike areas where there is too little rainfall to maintain a thick growth of vegetation. When the rain does come, it is often in raging torrents that wash away the sparsely vegetated soil and create the dramatic canyons and badlands of the American West. Elsewhere, however, erosion is unnatural, the result of man's misuse of the land.

On the whole, I feel that erosion control was the most important work I did at Redwood Park. In fact, I have an almost missionary zeal whenever I think of erosion control. But there is one thing I should not gloss over. Fighting erosion is a hard, heavy battle; and, as with any other worthwhile battle, there's a good chance that *you* will lose. Water erosion is a strong, persistent enemy. It's a fascinating enemy too: crafty, treacherous, sneaky, unforgiving, unforgetting, mindless, and merciless. Supposedly you can make a pact with the devil, but not with erosion.

In this chapter there are instructions for building check dams, contour trenches, and wattles. Follow these instructions and you'll have good reason to expect success. Most of the time. But there is also a good chance that an exceptionally heavy rain, exceptionally unstable soil, or a minor fault in construction will allow the water to wash your structure right away. When that happens, what are you left with? If you and whoever works with you did not enjoy the experience of working together, you are left with nothing. Less than nothing! But if the experience of building and planting was warm, cooperative, compassionate, and friendly, the project was a success whether the check dams hold or not.

As an engineering venture, you should build your structures as if they were going to last forever. Perhaps they will. But as a spiritual venture you should treat the whole thing as if success or failure of the structures is totally irrelevant. Make sure the process is human and loving, have fun, and open your eyes to the here and now. Saving soil is important, but not at the expense of losing a group of kids or a group of friends.

HOW EROSION HAPPENS

In the following sections I tell you what deeds you must do to fight erosion. But before you put on your coat of armor and rush out of the house, let's stop for a minute to examine the nature of the beast. Here is a model of a typically eroding watershed.

To begin at the beginning, drops of rain fall down. Plip, plip, plip. They hit the ground at a speed of about thirty feet a second. If your land is healthy and the raindrops fall onto a thickly carpeted meadow, a wonderful thing happens. It is something you have to see to appreciate fully. The next time it begins to rain, try to forget everything your mother taught you about "catching your death of cold," lie down on your belly, nestle your chin into the grass, and get a frog's-eye view of how raindrops fall. You'll see how the raindrops hit the individual blades of grass, causing them to bend down. This bending absorbs the energy of the raindrop, and the raindrop slides gently off the blade of grass, which immediately springs up again, waiting to catch another raindrop. Perhaps it's just my own sense of humor, but the sight of hundreds of blades of grass bowing down and popping back up like piano keys strikes me as one of the merriest sights in the world, and I've spent embarrassing amounts of time rolling around on wet meadows in the rain, laughing at the wonderful antics of the blades of grass.

After the energy of the raindrop is taken up by the grass, the raindrop slides gently to the ground. On a healthy meadow with lots of humus, the ground is spongy and absorbent and the raindrop quickly sinks out of sight.

A similar thing happens in a forest. As every kid knows, the best place to run when a sudden rain comes is under a tree — unless, of course, there is thunder and lightning. The leaves of the tree break the raindrops into a fine mist. What mositure does fall through the canopy is easily absorbed by the understory, the leaf litter, and the humus, and it too sinks gently into the ground.

But let's say that the ground has been logged, grazed, burned, cultivated, or otherwise disturbed. There are now bare patches of earth. When the raindrops hit a bare spot, they strike full force, like tiny hammers, and splatter the soil. This splattering breaks up clods of earth into fine particles. The raindrops hold the fine particles in suspension. As the water sinks into the soil, these fine particles get filtered out and soon clog up and seal the passageways through which the water would ordinarily flow. The clogging and sealing effect is very important: clear water percolates through the soil ten times faster than muddy water. After a brief time the soil becomes crusty and impenetrable, and the water can no longer sink in. Instead it forms puddles on the surface.

On flat land, the puddles loiter around, grow bigger, and form temporary ponds. The soil structure is damaged somewhat, but there is no real erosion.

On slopes, however, the water flows downhill over the surface of the ground, evenly, like a sheet. It carries dirt particles dislodged from the tops of hills and deposits them below, creating what is known as *sheet erosion.*

Probably the worst thing that can happen at this point is that the flow of water becomes channelized, either because of the topography of the land or because of an accidental occurrence like a furrow, a tire rut, or a cow path running downhill. The water gathers speed and the particles of dirt act like sandpaper. The water soon cuts a small trench or rill, which it may eventually widen and deepen into a gully.

As you can see, a gully is really the result of erosion — not the cause. Yet once the gully gets established, it brings about many severe problems. With each rainstorm, it gets deeper and deeper until it may even cut below the level of the ground water, draining it and lowering the water table.

We now have the beginning of a vicious cycle. As you probably

water table

gully

① ② ③ ④ ⑤ ⑥

a gully can eventually lower the water table

know, much deep-rooted vegetation depends more on ground water than on surface water from the rain. As the water table is lowered — both from lack of rainwater penetration and from the draining action of gullies — vegetation over the watershed becomes more meager and scruffier. In some places fields of thick grasses are replaced entirely by sagebrush and chaparral, with scraggly growth and much exposed soil. Less ground water leads to scruffy vegetation, which leads to more bare soil, which leads to more splatter, more soil clogging, less water penetration, more runoff, and a further deepening of the gully. As the gully deepens, it drains the water table still more, producing a further loss of vegetation, more exposed soil, more splatter, and so on for another downward cycle.

Meanwhile, as the gully gets deeper, the earth along its banks begins to cave in. Soon the gully sends out fingers that spread over the meadow, eating steadily away at the soil.

Within a few years, thousands of tons of topsoil are washed away, along with thousands of tons of subsoil. Where does it all go? Eventually, the gully probably drains into a stream. On a healthy watershed, a good cover of vegetation absorbs water, holds it like a sponge, and releases it gradually into the stream. The stream runs steadily and cleanly. But on an eroding watershed, the water runs off the surface with a heavy load of suspended silt, swoops through the gullies, and flushes out into the stream after every rainstorm. Instead of a clear, even-flowing stream, there is now an intermittent dry creek given over to flash floods. The silt kills whatever life there is in the stream and acts like sandpaper to cut into the stream bed and banks, causing further damage.

Sound dismal? It is! Yet this is exactly what is happening to thousands of small watersheds around the country. You should be aware of this process, but if your land has a few gullies, please don't get totally freaked out. Gullies can be stopped, and even if the "vicious cycle" has begun, there is a lot you can do to reverse it. That's what this chapter is about: how you can stop erosion without a lot of money, bulldozers, or a detachment from the Corps of Engineers.

So far I've given you a model of a typically eroding watershed, a model which should help you to conceptualize what is happening on your own land. If all you are going to do is think about erosion, you can stop here. But if you are going to *do* something about it, you will need a gut-level feeling for how erosion is happening on your land. This feeling, more than anything you read, will tell you where to plant, where to mulch, where to build check dams, and where to stay out of the way. It will prevent you from building a matchstick structure to stop a raging torrent, and it will save you the trouble of building a Hoover Dam to control a trickle.

In short, you've got to get wet! You've got to go out in the rain, lie belly down on your meadows, squish soil and mud through your fingers, look at the color of your water, and poke at the sides of your gullies to see how solid they are. Water is amazing stuff, and to see what it does, you've got to get intimately acquainted with it. Intimately! Right through to your socks.

A firsthand understanding of how your land is (or isn't) eroding will have its side benefits. It will get you out in the rain, which is sort of magical in its own right. It will also give you an appreciation for the strength, determination, and beauty of the erosion process. If you are going to fight erosion, it's much better to fight a beautiful enemy that you admire rather than an ugly enemy you hardly know.

One more thing. The Soil Conservation Service is an excellent ally in fighting erosion. You can find them by looking in the phone book under U.S. Government, Department of Agriculture. Personally, I think the Soil Conservation Service is one of the few good things the government is now doing. I've called upon them for various meadow, forest, stream, and gully problems. They have sent me (free!) grassland experts, stream experts, and soil engineers — persons who knew their subject well and who not only gave me advice but usable advice at that. My own good experiences with the Soil Conservation Service may have been accidental, but by all means give them a try.

FIGHTING EROSION WITH PLANTS

Of structures and plants Later on in this chaper, I'll explain
how to build structures that will stop erosion and hold your soil
together. Building these structures can be fun, almost like playing
with an oversized erector set, but please don't get hung up on them.
The Army Corps of Engineers seems to view erosion-control
structures as monuments, and in many places their cement bulwarks
are even more prominent and obtrusive than the original erosion.
Don't make that mistake. The structures I recommend are merely
temporary, even rinky-dink, devices to hold the soil together until
a permanent vegetative cover can get established.

The only successful and lasting way to fight erosion is with
plants. One of the nicest things about using plants is that plants
want to fight erosion. In fact, they want to fight erosion even
more than you do, and what's more, they know how to do it.
Take a blade of grass, for example. Grass depends for its
survival upon topsoil, and over the last several million years it
has developed ways of holding on to and increasing the earth's
supply of topsoil. Grass intercepts raindrops; it forms a tough,
tangled mat that prevents raindrops from flowing downhill; its
fibrous roots embrace the soil and hold it together. Decaying
roots create passageways through which water can penetrate, while
transpiration allows the grass to pump water out of the soil before
the soil gets waterlogged. At the end of its life, grass falls to the
ground, decays, and becomes humus, which is the best of all
possible elements in the topsoil. Plants depend upon a healthy
soil and they have learned how to serve and preserve that soil.
Every time you drop a seed into the ground, you are introducing
an ally with millions of years of genetic experience in fighting erosion
and tremendous willingness to put that experience to use.

Temporary cover The first thing you should think about when
you are faced with an erosion problem is a temporary (or emergency)

cover. You will eventually want to plant a permanent cover of native plants that will perpetuate themselves and restore the soil. But if you have a lot of bare land and an immediate danger of erosion, you have to act fast. You need some sort of temporary vegetation just to hold things together until the permanent vegetation can get established.

There are certain plants that have a special capacity for stopping erosion. I wish I could tell you exactly what you should plant on your land — I know it would make your life easier — but I can't. There are too many variables. I know that in the hills above Oakland I can get good results with a mixture of rye, barley, trefoil, mustard, and a few other minor flowers. But I doubt if this information will help you if you are in Indiana, Georgia, Vermont, or Alaska.

All I can do is give you some general advice about what you should look for in an emergency cover plant. For the specifics you'll need local guidance from your Soil Conservation Service, county agricultural extension agent, your local hermit and organic gardener, or your local seed dealer (who often has a special "erosion-control mix"). Or you can look into some of the books I recommend at the end of this chapter which give a species-by-species run-down of many valuable erosion-control plants and tell where they can be used.

The ideal erosion stopper is a plant that:

Germinates quickly and easily;

Grows fast *before* the first heavy rains;

Has a dense, fibrous root system;

Is frost resistant;

Is temporary — make sure the recommended "wonder plant" isn't some horrendous weed that will take over everything in sight;

Is a mixture. Don't depend on one plant, no matter how good its reputation. And make certain that at least one element of the mixture is a legume (member of the pea family). Legumes do for the soil what yogurt does for the intestines — they foster lots of beneficial microorganisms that do much of the real heroics in creating healthy soil.

How to plant a temporary cover The best way to establish a temporary cover is first to dress the ground with a light sprinkling of very well-rotted manure or compost. You might want to work it into the soil a little bit with a hoe and then rake it some — but not too deeply, please. If you have erosion, you want to disturb the soil as little as possible. Once you've prepared the soil, simply broadcast the seeds a day or two before you expect rain.

Fertilizers and exotics What if you are dealing with a huge area, or if you don't have enough manure or compost for even a small area? Here's what you do. Scratch the surface of the ground slightly with a rake. Then spread the seed before you expect a rain. Wait until the seed has germinated and growth is under way, then carefully add an appropriate chemical fertilizer. (The Soil Conservation Service or a local seed dealer will tell you how much seed to scatter and what kind of fertilizer is "appropriate.")

Aside from chemical fertilizer, there is another bitter pill you may have to swallow. Some of the most effective plants for erosion control are exotic grasses and clovers. Call me a native-plant chauvinist, but I normally abhor foreign exotics. I have very high standards about not using them. For that matter, I have very high standards about not forcing growth with chemical fertilizers. Yet when the soil is bare and the rains are due, I am faced with a clear choice: I can either hang on to my standards, or I can hang on to my topsoil. Standards can be replaced, rationalized, or even forgotten within a week. Topsoil takes thousands of years to form. Whenever I've had to make a choice, I've opted in favor of topsoil.

If you do decide to use an exotic, there are special guidelines you should follow. Make sure the exotic has been around for a long time and is well tested in your area. Make especially certain that it won't escape and spread all over the place.

After seeding   If your land is relatively flat, you can seed and forget. But what if you're working on a steep slope where the soil is so unstable that you're afraid it will wash away, or where the land is so hard that you think the seeds might simply float down the hill? In such cases you'll have to devise some way of holding the seeds and earth in place — at least until the seeds germinate, roots work their way into the soil, and green stuff rises up like flags of victory to tell you everything is going well.

Willow stakes   In the following sections I describe several structures that will hold the soil together for a while. You can use any materials to build these structures, but if you use willow cuttings (see pages 110-14), you will reap an extraordinary advantage. Not only will they serve a mundane mechanical function as posts or stakes, but they will very likely sprout, send down roots, help bind the soil, and carry on an exuberant and useful existence of their own. Willows are especially valuable wherever you're dealing with moist land and bad drainage.

In addition to willows, there are other cuttings you can use for living stakes or posts. In our part of California, for example, elderberries and "mule fat" sprout easily from cuttings. Under hard conditions they may last for only one or two seasons — but while they last they'll do a lot of good.

Black locusts   The black locust is not an insect; it's a tree with a supergood reputation for erosion control. It establishes itself on poor, dry sites, has a spectacular rate of growth and a good root structure, and adds a lot of nitrogen to depleted soil. It is not unusual for a three-year-old locust to be fifteen feet tall (thank goodness it's not an insect!) with a root system spreading twenty-five feet.

You can plant locusts as seedlings or from root cuttings (see pages

105-8). For erosion control, plant them close together — say, five feet by five feet, or even three feet by three feet in really bad places.

Permanent vegetation   Temporary vegetation is meant to give out, and even willows and locusts are not usually climax species. You should plan for what you hope the permanent vegetation will be. Talk it over with your land. Find out what was there before the land was misused. Decide whether the land can support its climax vegetation, or whether you should begin further down the line of succession. I can't advise you what to plant — it varies from one area to another, and in fact from one acre to the next — but by studying uneroded, undisturbed land in your neighborhood, you should be able to figure it out.

The best time to plant permanent vegetation is just as soon as the temporary vegetation has stabilized things — usually toward the end of the first rainy season.

CONQUERING THE SPLATTER

It may sound silly and quixotic to you, but if you are going to control erosion, you must begin by fighting raindrops. Raindrops hammer insistently at your land, and to prevent damage there are two things you must do. First, you've got to make sure there is something waiting to intercept the raindrops before they hit bare soil: vegetation, if possible, or some sort of mulch. Secondly, once the raindrops fall, you've got to stop them, corral them, and let them sink into the ground. If, perhaps with trenches, brush mats, or wattles, you can get the raindrops to sink into the ground wherever they fall, there will be no runoff, and thus no erosion.

How to recognize sheet erosion   Sheet erosion, according to the people who measure such things, causes 80 per cent of all topsoil losses. Gullies cause only about 20 per cent. Yet a gully stands out like a wound, screaming for attention, while sheet erosion happens so gradually, almost invisibly, that it's hard to detect. You think everything is all right until one day you wake up and realize that your topsoil is gone. Sheet erosion is very insidious.

Is your land suffering from sheet erosion? Looking for sheet erosion is a little bit like searching for a snake. If you merely walk around, sniffing flowers and lackadaisically enjoying whatever strikes your eye, you are unlikely to see a snake. But if you make a special effort to find one, turning over logs and stones, looking hard between the blades of grass and around bushes, you will probably find several snakes in a few hours.

The same is true of sheet erosion. You have to go out into your fields with nothing else on your mind except looking for sheet erosion. Don't get waylaid by flowers, butterflies, or ripe strawberries. Keep your mind on your task. Climb to the top of a hill, forget about the view, and look down at the soil. Here is what you should be looking for.

BALD SPOTS on the hilltops and slopes, often with a build-up of fertile soil down below.

EXPOSED ROOTS   Roots of trees, shrubs, and other plants do not *grow* out of the ground. If the roots are exposed, it is because the soil has been washed away.

STAINS ON OLD FENCE POSTS   These sometimes show that the soil was once deeper than it now is.

EXPOSED ROCK   If you feel that your meadows have been getting rockier and rockier each year, unless your land is a gathering spot for meteorites, this is a sign that the soil is being washed away.

Mulch   Once you discover sheet erosion, don't waste too much time either admiring it or bemoaning it. Get the right mixture of seeds, put them in a wide, shallow basket, and go skipping across your meadows like Ceres strewing the seed. Be joyful — and the seed, the land, and perhaps the universe will respond to your joy.

In most places you can seed and forget. But if the soil is loose and unstable, or if it is so hard that you're afraid the seed will wash off, or if the slope is exceptionally steep, you should apply a mulch after you've seeded. A light covering of mulch does wonders. It cushions the impact of the raindrops, like those blades of grass, and allows the water to settle in gradually. It creates a network of little dams on the ground that impound the water and prevent it from getting a running start down the hill. It absorbs water. And as it decays, it adds organic matter that eroding land usually needs so desperately.

When you mulch, follow the suggestions on pages 42-46. But remember this important difference: underneath the mulch are seeds, and you want to encourage, not smother, their growth. So keep the mulch covering thin — no more than an inch or two — and avoid any mulch that tends to mat down.

Straw is far and away the best mulch you can get for erosion control. But don't be too fussy; other mulches also work very well.

Brush mats Brush mats are for really nasty places — places where you want to use a mulch but where the slope is so steep that you're afraid a loose mulch will wash down the hill. Believe me, an eroding hillside with a huge pile of soggy mulch at its base is a nightmarish sight. The way to avoid it is to use brush as your mulch and tie the brush together into mats.

brush mat

To make a brush mat, first lay two wires parallel to each other on the ground, about two feet apart. Lay the brush over the wire. If you use fir boughs or pine boughs, pile them very thin; otherwise, they'll smother the seed. If you use sparser chaparral brush, you can make the mats as much as six inches thick.

After you arrange the brush over the wires, bring the wires back over the top of the brush. Use baling wire to connect the upper and lower strands of wire. Pull them tightly together and tie them off, making a connection every six inches or so. The loose ends can be twisted tight with pliers.

You now have a brush mat that will hold together very effectively, even on quite steep slopes. If you want to be extra safe, you can stake your brush mats down to the ground — preferably with sproutable, rootable stakes.

Contour trenches   Here is still another technique you can use in addition to mulching. If by some chance you don't have any mulch, you can sometimes use this technique instead of mulching.

Contour trenches are simply ditches that you dig along a hillside, following a contour and running perpendicular to the flow of water. They catch water and allow it to sink into the ground before it can get a running start down the hill. Contour trenches are particularly valuable on hardened soil — like old logging roads — where water penetration is painfully slow.

To make contour trenches, first gather all your friends and issue them picks, mattocks, and shovels. When the moaning and groaning stop, begin digging several short trenches five or six inches deep and no more than about two or three feet apart. Keep this project short! Digging ditches on a hard-packed, heavily eroded slope is nobody's idea of great fun.

Brush wattles   Simple seeding, mulching, brush mats, and contour trenches will take care of 98 per cent of your sheet erosion problems. For those rare times when you have an exquisitely nasty and persistent problem with sheet erosion, you can resort to brush wattles.

Begin by making a series of contour trenches at least eight inches deep, preferably deeper. As you remove the dirt, somehow, somewhere, get it out of the area. Next, lay some brush in the trenches. Stagger the brush along the trench so that it all interlocks, like strands within a rope. As you build up the brush, stomp it hard so that it packs into the trenches. If it keeps springing up, you can try cursing it or packing it down with some dirt. The last several pieces of brush that you lay in the trench should stick up above the level of the land. To help keep the brush in place, knock in stakes (preferably stakes capable of growing) just behind the trench on the downhill side. Space the stakes one foot, or at most two feet, apart. If you have lots of long, limber branches, you should weave them between the stakes to form a wattle fence.

What you're left with is admittedly a weird structure, and one

Contour trenches

wattle fence

that is hard to build — especially on a steep, unstable slope where you are most likely to need it. It has, in fact, only one redeeming feature: it works! The water running downhill sinks into the trenches. Silt suspended in the water also gets caught in the trenches and builds up within the protruding branches of the brush and behind the wattle fence. A wattled slope soon forms little terraces of relatively stable silty soil — excellent places for plants to get a start.

GULLIES

Patrick Henry (of "liberty or death" fame) once said, "Since the achievement of our independence, he is the greatest patriot who stops the most gullies." I used to think this statement a bit outlandish, but the more I've gotten to know about land, gullies, and patriotism, the more I've come to agree.

Rills   The easiest way of stopping a gully is to catch it early. Whenever you see small rills (or channels), get right to work. Use a mattock or a hoe to break them up. Work in some compost or rotted manure, if you can, and rake the area smooth. Then treat the area as you would for sheet erosion — seed it, mulch it, or possibly use brush mats or contour trenches.

Gully monsters   A neglected rill may grow up to be a monster gully. In the next several pages I'll tell you how to go about fighting and conquering gullies. It's a long, complicated fight, but very much worth the trouble. We no longer have fire-eating dragons, but we do have land-eating gullies to fight. Just to make sure you can find your way through the following instructions, here is an outline of the battle plans.

1. Stabilize the gully bottom. The bottom is more important than the sides. If the gully continues to dig deeper, no matter what else you do, the sides will cave and slump. You've got to prevent the gully from getting any deeper, and you should even attempt to build up the bottom.

2. Grade the walls of the gully to their *angle of repose*—the angle at which they will no longer slump or slide.

3. Stop or reduce the flow of water entering the gully.

4. Plant an immediate cover of grasses and legumes that will hold everything together for a season or two.

5. Plant a permanent cover of native shrubs, trees, vines, and grasses that will eventually stabilize the area, perpetuate themselves, build up soil fertility, encourage wildlife, and completely restore the land.

Check dams   The way to stabilize the gully bottom and build it up again is with check dams. Please don't be intimidated by the thought of building a dam. You're not going to be competing with Grand Coulee or Aswan. In fact, your check dams won't even hold any water. They are merely obstructions that will slow the water down. And the best of all possible obstructions (as we all know from our various misadventures in life) is a big mess. Basically that is what a check dam is: a big mess of brush or perhaps straw packed into the bottom of the gully, with a simple structure to hold it all in place.

Why a check dam works   I think we all have an intuitive sense of why a check dam works: a slow-moving stream carries far less silt and does far less damage than a raging torrent. But to understand how dramatically true this is, you might want to consider a few hard-core engineering facts. If you reduce the speed of the flow of water by one-half, here (according to certain laws of hydraulics) is what happens.

The erosive or cutting capacity of the water is reduced about four times.

The quantity of silt that can be carried is reduced about thirty-two times.

The size of particle that can be transported by pushing or rolling is reduced about sixty-four times.

One year later

As you can see, by slowing down the flow of water, you reduce the amount of damage it can do, and you very spectacularly reduce the amount of silt it can carry. If there is lots of silt suspended in the water, once you slow the water down, most of the silt will be dropped — thus building up the bottom of the gully again.

The principles of check dam architecture There are many possible designs and materials for building check dams, but whichever one you choose must adhere to certain architectural principles of check dam construction.

HEAD-TO-TOE ALIGNMENT   The most effective way of building check dams is to build them in a series where the base of the upper dam is on a level with the top of the lower dam. This will eventually stabilize the whole gully bottom and will create a series of steps or terraces.

SMALLNESS   "The bigger they are, the harder they fall" applies particularly to check dams. For most gullies, the check dams should be no more than about two feet high. Anything much higher than two feet will necessitate anchors, *deadmen*, and other retaining-wall features. Several small dams are far more effective and easier to build than a few big dams.

DIGGING IT IN   The dam must be dug into the walls of the gully, not just laid genteelly up against them. Unless the dams are dug far enough in, water will sweep around them.

NOTCHING   A notch is a place where the water can flow over the dam. This is essential. Without one, the silt builds up behind the dam, the water flows on top of the silt, and instead of being led through the notch, it may start eating away at one of the slopes. Eventually, it may make a new channel around the dam. I've seen many erosion-control dams standing proudly and nobly on dry land while gullies flowed merrily around them.

APRON   Once the silt builds up behind the dam, the water flows through the notch like a waterfall. You'll need an apron to catch it before it digs out a pool and undermines the dam. The easiest apron is a bed of stones where the water can simply knock itself out and flow tamely to the next check dam.

Building a check dam   There are several possibilities for building very good check dams: a rock dam, a wire dam, a stake dam, a pole dam, and a plank or slab dam. Which one you choose to build will probably depend more upon the materials you can scrounge up than upon anything else. I built mostly pole check dams because we had plenty of poles. Whichever one you decide on, remember to follow the general principles already laid out, and you will make out very well.

head-to-toe alignment

Slab Dam

Wire Dam

Stake Dam

Grading the slopes After you build the check dams, your next step is to break down the steep gully walls to their *angle of repose.* To my ears, "angle of repose" is one of the most beautiful phrases in the language. Unfortunately, it's far easier to say it than to do it. I know of no easy way of breaking down steep, clifflike slopes. Professionals sometimes use dynamite and bulldozers, so I've been told, but all the bulldozer operators I've ever met are scared to death of working along the rim of a sizable gully. When it comes to grading gully slopes, the machine age has deserted you, my friend, and what you are left with, wonder of wonders, is your hands! So get together a collection of picks, mattocks, shovels, and digging bars, round up everyone you know who owes you a favor, and get on with it. Knock off the sharp edges, and wherever you can, gentle out the steep slopes.

As you are working, you'll be knocking tons of earth down into the gully bottom. The first rains will dissolve this earth, spread it out, and deposit it behind the check dams to raise the bottom. You can help this process along, and also prepare the bottom for planting, by breaking up whatever heavy clods fall into the bottom. If you have any water, you might also wet the dirt down to compact it and further ready it for planting.

Once the slopes have been graded to their angle of repose, you should treat them for sheet erosion, with seed, mulch, or the other devices recommended in the previous section.

Limiting the water flow You now have to make certain that as little water as possible enters the gully. Where is the water coming from that originally carved it out? You must find that water, even if it means going out in the middle of a rainstorm.

You can usually restrict the flow by treating the area above the gully head for sheet erosion. Contour trenches usually work quite well, and as a last resort brush wattles are nearly infallible. Whatever treatment you use, make sure you extend it far up the slope.

Occasionally an expert will appear in your life and suggest that you divert the flow of water away from the gully. He will urge you to build a "diversion ditch," perhaps with an "entrapment compound." He will probably pull out a pencil and paper and make a few fancy diagrams. When you meet such an expert, the first thing you should do is grimace, pound your chest, jump up and down, and point excitedly to the sky. If this doesn't scare him off, grab your hat and run. As you can guess, my own experience with "diversion" has been disastrous. Diversion does not solve any problem; it just moves the problem somewhere else.

Planting Once you've stabilized the bottom of the gully, graded the slopes, and reduced the flow of water, you have completed the mechanical aspects of controlling the gully.

Now you should plant. Use the previously mentioned routine of temporary planting followed by permanent planting. Don't plant anything in the bottom until the silt has collected into terraces. Then you can plant moisture-loving trees right in the silt, where they'll usually thrive.

Maintenance Remember the little Dutch boy who put his finger in the dike, held back the ocean, and became a culture hero to all five-year-olds? I don't suggest you spend all next spring with your finger in a check dam, but the Dutch-Boy Principle still holds: small leaks can be easily plugged. Sometimes all that is necessary is for you to shove a few pine boughs in at the right place. If you do, silt will continue to collect. If you don't, the leak will often get bigger and bigger, bringing the whole dam down. You should also check to see that the mulch is still in place, the grass has germinated well, and no heavy flow of water is entering the gully. Visit your check dams as often as you can during the first one or two seasons to see how well they are holding up and to solve minor problems before they grow.

Culverts Culverts are pipes that bring water under a road or trail. They are responsible for thousands of gullies in every state. Road engineers have a strange idea that if they install these culverts at a steep pitch, the water will throught them very fast and keep the culverts clean of debris. Road engineers really get turned on by "'self-cleaning" or "self-maintaining" culverts. But as I've already mentioned, the fast flow of water increases its erosive powers many times over. And often at the dump end of the culvert you will find a huge gully.

If there is already a gully, you have no choice but to go ahead with the gully trip. But if you can catch the problem early, the best thing you can do is dump a lot of rocks, broken asphalt, or cement rubble under where the culvert lets out. This will break the force of the water, acting much like an apron beneath a check dam. If you do this wherever you have a culvert, you will save a lot of aggravation and a lot of soil as well.

Afterwards I don't want to minimize the fact that controlling a gully is hard work. But it is necessary work, and in the long run extremely satisfying.

Once you have brought a gully under control, watch it closely and uncritically. You may be in for a surprise. Some of the most beautiful places I know are old, stabilized gullies. When you are fighting a gully, you are primarily fighting erosion damage. But you are also creating a shady, potentially lovely, miniature canyon which will collect moisture, support many plants, and become a wonderful refuge for wildlife. Turning a barren gully into a lush pocket of life is the nearest a human being can come to an oyster, which turns its injuries into pearls.

STREAMS

If you have erosion problems along your stream, keep in mind that these problems are almost invariably the end result of sheet erosion, gullying, faulty road construction, overgrazing, and other abuses throughout the watershed. This section will tell you a few things that you can do to help alleviate the sufferings of an eroding stream or creek. Do what you can, but always remember that working directly on a stream is only a stopgap measure. The ultimate healing must take place throughout your whole watershed. If you ignore this advice and concentrate your efforts on the stream, your watershed will continue to go downhill — literally — and each year you will find yourself having to build bigger and uglier structures in your stream, stonewalling its banks, sandbagging, jettying, deflecting, and diverting, until eventually (with the best of intentions) you have converted a living stream into a grotesque plumbing problem. In the end you will hate your stream for being so ungrateful to you, and your stream will hate you for being such an oppressor. So avoid the Army Corps of Engineers nightmare. Remember that a living stream — like all living things — is bound to be rebellious. Respect its rebelliousness. Do a few light projects to help your stream along, but if the stream doesn't respond, don't force it. You are trying to assist your stream, not defeat it. Keep your stream projects modest, and by all means concentrate most of your erosion-control efforts on the watershed around the stream.

Bank erosion   Streams can erode in two directions: they can dig increasingly deeper channels for themselves (channel erosion), or they can eat away at the banks. If your problem is bank erosion, there are several steps you might take.

First of all, stop all physical injuries to the banks. In particular, stop grazing animals (cows, horses, and sheep) from breaking down the banks to get to the water. You may have to fence off parts of the stream and, if necessary, even build a watering trough away from the stream's edge.

Next, you can build deflectors. Deflectors are basically piles of stone placed upstream from an eroding bank to absorb the force of the water. But you can't just dump some stones into the stream and hope they'll do the job. You've got to build a deflector as carefully as you would any other structure. First, dig some of the stones into the bank; otherwise the water will nibble, nibble, nibble, all day and all night, never resting until it eats its way around the stones and eventually leaves them stranded midstream as in ineffectual island. Next, lay the other stones out from the bank a couple of feet into the channel. Mortar the stones together if the stream is dry. Or if the stones are

rock deflector

all sizable, you can move them around so that they are stable, using smaller stones to chink the cracks. Or, if all you have are smaller stones the size of footballs, you can try corralling them with logs dug into the bank.

Finally, you should plant the banks heavily. Willows, planted as stakes along the banks, are particularly good. (See pages 110-14.)

Channel erosion   Another serious problem with an eroding stream, similar to that encountered in gullies, is that it often cuts deeper and deeper into its channel, causing the banks to slump and eventually draining and lowering the water table.

I once spent a lot of time building well-crafted pole check dams in the streams of Redwood Park. The idea was to slow down the water, get the silt to build up in the stream bed, and raise the water level of the creek. It sounded good to me and to the several erosion-control engineers I talked with. The stream, however, had different ideas. The first rainy winter it rose up, flexed its muscles, and knocked down every one of my well-crafted check dams.

In gullies, of course, pole check dams work quite well. In a stream they also work very well for a season or two, collecting big and impressive baskets of silt. But it doesn't do a bit of good. You can't get trees and grasses to grow in the middle of a stream as you can in a gully, and most assuredly in two, three, four, or five years, your dams will collapse and their fine collection of silt will be washed downstream.

While pole check dams do not hold up, the hydraulic principles are still valid; namely, as you slow down the flow of water, its erosive power is cut drastically. You can, if you want, construct permanent, well-made rock check dams. But I have come to prefer a more casual approach. Every year before the first fall rains, I'd gather an ax, a shovel, a strong bar, and a couple of strong kids, and we'd set out for a walk along the stream. When we came to rocks or fallen logs lying along the bank, we'd pry them into the stream. Then we'd arrange them into small, *temporary* check dams. We'd key a few rocks well into the bank (otherwise the water would sweep around the rocks and eat away at the bank), we'd make the rough equivalent of a notch to handle the overflow, and we'd arrange some stones beneath the notch to act as an apron.

Check dams such as these are very makeshift, guaranteed to last no more than a season or two. But you are not worried about permanence. You do not want future generations to gaze in wonder at your check dams. Make them small, tight, and frequent. What you are trying to do is break up a swiftly flowing stream into several pools, slowing the water down so that it will do less damage to the stream bed. Every year you will have to rebuild your makeshift dams, but that's all right, since the permanent cure for your stream's problem lies elsewhere, throughout the watershed. Until then, like an earth doctor, I suggest you give your stream not only annual check-ups but annual check dams — dozens of little stone check dams to help it along until the watershed regains its good health.

READING

In other areas of conservation there is pitifully little information. Not so with erosion control. The 1930s were dust bowl years, gully years, and Civilian Conservation Corps years. The CCC, the Forest Service, and the Soil Conservation Service all published loads of erosion-control pamphlets and books. Every field worker who developed a new style of check dam — and there were hundreds — published a description of it. Sometimes the check dams collapsed within a few years, but the publications live on to clog our minds. The problem I've had with erosion-control literature is wading through it all for what seems sound, relevant, trustworthy, and useful. Here are some of the books I have found especially handy for small-scale erosion-control projects.

Handbook of Erosion Control in Mountain Meadows, by Charles J. Kraebel and Arthur F. Pillsbury. California Forest and Range Experimental Station: U.S. Forest Service, 1934.

For most people this is probably an impossible book to get hold of, but by all means try your best. It's the most thoroughly practical book I know, with lots of simple suggestions for controlling gullies. There are excellent diagrams and a strong emphasis on using native materials.

A Study of Early Gully-Control Structures in the Colorado Front Range, by Burchard H. Heede. Paper No. 55. Rocky Mountain Forest and Range Experiment Station: U.S. Forest Service, 1960.

This publication is a review of several Civilian Conservation Corps structures, examined twenty-five years after they were built. It shows which ones stood up, which ones failed, how they failed, and why they failed. It's very instructive. Here is your chance to learn from someone else's mistakes.

Grass in Soil Erosion Control, by Layman Carrier. SCS-TP-4. Washington, D.C.: Conservation Service, 1936.

This pamphlet gives a short list of various grasses and discusses their erosion-fighting values.

Results of and Recommendations for Seeding Grasses and Legumes on TVA-CCC Erosion Control Projects, by J. H. Nicholson and John E. Snyder. Norris, Tennessee: Tennessee Valley Authority, 1938.

This list of grasses and legumes rates them according to where they will grow, what their moisture and soil needs are, how well they bind the soil, and how well they build up soil fertility.

Trees and Shrubs for Erosion Control in Southern California Mountains, by Jerome S. Horton. California Forest and Range Experiment Station: U.S. Forest Service, 1949.

Giving a plant-by-plant list of several species of tree and bush, this book tells where to plant them, when to plant them, and even how to plant them. It also has detailed diagrams of various erosion-control structures. It's too bad this valuable book is so limited in geographical area. You might check to see if your own Forest and Range Experiment Station has a similar publication.

The Stream Conservation Handbook, edited by Nathanial P. Reed. New York: Crown Publishers, 1974.

This book claims that "the primary objective of stream improvement is the restoration and enhancement of trout habitat." It was written for fishermen, many of whom are beginning to band together into groups like Trout Unlimited to maintain their streams. The big-stream scale of this book will probably make it not very handy for small landholders. But if you do happen to have a fishing creek, it will tell you what you have to know to keep it fishable.

6 *The Seed Bag*

During the month or two when most California plants go to seed, I would wander over the meadows collecting bags and bags of seeds — wildflower seeds, tree seeds, grass seeds, shrub seeds, vine seeds, even thistle seeds. My seed collection was the envy of every squirrel and gopher in the state.

Not only did I collect seeds myself, but I also involved hundreds of kids in the activity. Sometimes the seeds we collected would get planted at once. Other times I'd extract them, dry them, store them, and treat them before giving them back to other kids to plant. It all sounds terribly complicated, but it isn't. I've been fairly successful without being a botanical genius. I think anyone else can be equally successful, even if he or she has never so much as put a radish seed in the ground. That's what this chapter is all about: a step-by-step procedure that will enable anyone — i.e., you — to begin collecting seeds and growing wild plants from them.

"But why bother?" you may ask. Don't wild plants take care of themselves — the less interference from us the better?

There are many frivolous reasons for collecting wild seeds. Who can resist trying to increase a stock of baby-blue-eyes or trilliums simply for the joy of it? But there are practical reasons too. In fact, the more seeds you have, the more uses you find for them. Seeds have proved indispensable to me in erosion-control work for revegetating gullies, washed-out roads, unstable slide areas, and on hillsides where the soil was getting dangerously thin. After a fire it was a pleasure to have several pounds of wildflower seeds to give to the charred soil as a sort of get-well present. I used seeds to help repair trail damage, to increase wildlife forage, to replace edible plants I'd been eating, and to landscape buildings with native plants. I'd spread the seeds of certain lupines, clovers, and vetches in places where I wanted to improve the soil's fertility; I'd carefully collect and treat the seeds of rare plants to help ensure their survival; and I was always trading seeds with other wild-seed buffs. Trading seeds turned out to be unexpected fun. How much can *you* wheel and deal for a half-pound sack of fawn lily seeds?

You can even earn money collecting seeds. You won't get rich, not by any means, but wild plant seeds are definitely salable. Try local garden clubs, native-plant societies, highway departments, or dealers. (For a partial list of wild-seed dealers, write to the National Arboretum, Washington, D.C.)

But while there's a chance you might sell a few seeds, I assume that you're far more likely to be collecting seeds for your own use. Let me give you one piece of advice (which you'll probably ignore): don't overcollect! Collecting seeds can be addicting: once you start, it's hard to stop. Collecting is fun, but planting seeds is hard work. You can't just scatter them lovingly over the earth. If that's all you're going to do, why not stay home and let nature do the job for you? If you really want to grow native plants, you may have to work the ground like a garden. You may even be wise to germinate the seeds in flats, transplant the young plants into pots or cans, and later (sometimes much later) plant them out in the field. Bear this in mind, and limit your collecting to what you can handle.

For me the whole seed experience has been an aesthetic and spiritual adventure of the first order. I used to think that the flower season was over once the soft, fragrant petals dropped. Now I keep my eyes open and watch the falling petals give way to another kind of beauty — the geometrical, architectural beauty of seed pods and seeds.

Throughout the hot, dry, withering California summer — when the flowers were gone and everything in nature seemed to have died — I'd go out with groups of kids to collect these seeds, treat them, watch over them, and think about the life contained within them.

Seeds are very beautiful and very mysterious. In the following sections I describe the specific techniques for collecting, treating, and planting them. It's certainly important to understand these techniques, and by using them I'm sure you'll have great success in getting wild plants to grow. But there is something beyond success. The longer I have been dealing with seeds, the more I have gotten fleeting perceptions of their beauty, their power, and their mystery. Over the years, as my intimacy with seeds has increased, these fleeting, flickering perceptions have been growing stronger and steadier — until I have come to value these expanding perceptions almost as much as I value the thousands of wildflowers I have grown.

COLLECTING AND EXTRACTING SEEDS

Learning to recognize wild seeds You may feel that you don't know very much about wild plants, you hardly know one wildflower from another, and you haven't the foggiest idea what their seeds look like. Join the crowd. There are 20,000 wild plants in the United States, and everybody — even the expert who drops Latin names — feels somewhat inadequate to the task.

Yet there are several dozen seeds that you do know, even now. You know acorns, chestnuts, maple wings, and pine cones. You probably know rose hips, several berries, May apples, sunflower seeds, milkweed seeds, and lots more. You undoubtedly know enough to get started right away.

As for learning more about seeds, there's only one way. You've got to go out and observe. It's easy and it's fun. Fix your attention on a flower (stake it out if necessary), return every few days, and you will see the seeds develop, ripen, and eventually scatter. Or when the plant is flowering, concentrate on its leaves, which will identify it after the flowers have fallen away. You'll be amazed to find how common flowers you've always vaguely known produce those barbed and corkscrewy seeds you've been pulling out of your socks since childhood. You'll also come to appreciate, I think, that in their own way seeds and seed pods are as beautiful as the flowers they come from.

When to collect seeds   The best time to collect seeds is when they're dry and ripe. A ripe seed no longer depends on the mother plant for nourishment, and it's ready to enter the world — "a little plant in a box with its lunch."

To test for ripeness, crush a seed. If it gives off a milky or gelatinous substance, it's not yet ripe.

How to collect seeds   Big seeds (especially nuts) can often be collected directly off the ground as long as they're not wet, moldy, or wormy. But most seeds must be collected while the pod or capsule is still attached to the plant. The idea is to collect the seeds just as the capsules are beginning to open but before the wind steals the seed — a period often lasting several weeks. For most flower seeds, you simply run your hand up the stem, using your fingers as a comb, to collect both seeds and capsules. Be sloppy! Let lots of seeds get away. You don't want to strip an area completely.

For grass seed you might imitate the Indians, who grabbed a handful of stalks, bent them over a basket, and shook out the seeds.

Although it's best to collect ripe seeds, there are times when this is tedious. Some plants, like mustard, ripen in such a way that on any one stalk there are dozens of green pods to every ripe pod. In such cases you should wait until the pods or capsules are as nearly ripe as possible. Then pick the entire seed stalks, lay them out on newspapers in the sun, and wait. In a few days most of the pods will have opened, and you can shake the seeds out. Or, as an alternative, you can collect the whole stalks, put them upside down in paper bags, hang the bags in a dry, warm place, and the seeds will separate from the pods.

There are a few seeds that *must* be collected when the pods are green. Certain lupines and vetches, for example, do not have seed pods which open gradually. The pods explode — bang! — and shoot the seeds considerable distances.

For plants like these, collect the seed pods when they are still green, with a section of stem attached to each pod. (The unripened seeds must still draw nutrients from the stem.) Lay the seed pods out on newspapers in the sun and cover them with a layer of cheesecloth or screen. After several days the sounds of battle will fade, the cheesecloth will stop jumping, and the seed is ready to be collected.

Labeling   I generally collect seeds in paper bags, and as I collect I write relevant information right on the bag. Here's an example:

> *Species:* Layia platyglossa (Tidytips)
> *Date collected:* 6/4/74
> *Place:* West-facing slope of Pinehurst Knob
> *Environment;* Dry, rocky, well-drained soil;
>   open, grassy, sparse vegetation.

Proper labeling is *very* important. I know how corny and high-schoolish that sounds, but it's true, nevertheless. Nothing is more frustrating and wasteful than to come upon a bag of fascinating seeds, bursting with vitality, and not to know what they are. You don't know where to plant them, when to plant them, or how to treat them. Viable seed looks happy and prosperous, and to consign a bagful of happy, prosperous seed to a sure death is heartbreaking. Avoid this tragedy: label your seeds carefully!

A side benefit of labeling, by the way, is that it trains your observation and increases your knowledge of the environment. Certain of our flowers grow only on rocky, serpentine soil. I never would have discovered this if I hadn't had a blank space on a label to fill in. Now, wherever I find patches of these flowers, I kick the ground and, sure enough, I find serpentine rocks. By observing the placement of such flowers throughout the park, I'm beginning to get a picture of the geology of the area — something I find delightful since I'm totally intimidated by geology. But once you get into seed collecting, all sorts of intimate knowledge comes your way without your really asking for it.

Extracting   After you've collected the seeds, you'll have to extract them from their pods, capsules, fruits, cones, or whatever kind of container they come in. The purpose of extracting is to separate the seeds from other materials that might absorb moisture and cause rot. If you're at all compulsive or perfectionistic, this isn't the job for you. Without a ridiculous amount of effort or equipment, you'll *never* get the seeds perfectly clean. And when you throw away the chaff, pulp, or cones, you'll *always* be throwing away lots of perfectly good seeds. Reconcile yourself to waste, and look at it this way: nature often produces millions of seeds to get one plant that will survive. However wasteful you may be, you are many times more efficient than that.

For many seeds it's enough to crush the pods or capsules by hand, put everything in a bag, and shake the bag vigorously. The seeds fall to the bottom while the chaff remains on top.

If the bag-shaking procedure doesn't work out too well, you can play around with colanders, strainers, and screens of varying meshes. Or try tossing the seeds in the wind to see how well they'll winnow. (An electric fan is probably better than the wind for winnowing, but it's hardly as picturesque.)

Another trick for round or smooth seeds is to dump the mix onto a blanket and then tilt the blanket until the seeds roll or slide off, leaving behind the angular pieces of capsules.

For fleshy fruit like berries and rose hips, soak the fruit in warm (not hot!) water until you can crush the pulp. The seeds will sink to the bottom and the pulp will rise to the top. Dry the seeds at once, unless you're going to plant them immediately.

Cones   Pine cones, fir cones, and other cones present special

Douglas Fir
Cone and seed

problems in both collecting and extracting. The cones you find on the ground are usually open and the seed has long since fallen away — or been eaten by birds and squirrels. The fat, closed cones that are full of seeds are usually swinging 50, 100, or even 200 feet above your head.

Professional seed collectors sometimes use rifles to shoot the cones down from especially desirable trees, but I assume you'll find this too impractical or too noisy, you don't have a rifle, you don't have a hard hat, or you're a lousy shot. There are ways of getting around this. First off, keep your eyes out for squirrel caches in the corners of old buildings and elsewhere. (If you feel guilty about ripping off the squirrels — and you should! — leave them some store-bought nuts.) Also watch for logging operations nearby, or do your own thinning at a time when the cones are ready to be collected. Or keep an eye out for trees that grow on the edge of clearings. These trees often have low branches that may bear cones within reach. A final trick is to check paved roads. At certain times you can sweep the seeds off the road as they fall from the trees. Or you can sometimes collect the cones by standing on the roof of your car or truck, or by setting up a ladder from a truck bed. You should collect cones when the bracts are just opening or loosening their hold on the seed.

The next problem is to extract the seeds from the cones. Put the cones in a warm room, or in the sun, for several days or even a couple of weeks. The scales will often open and you can shake the seeds out.

If the scales don't open, you'll have to bake the cones in the oven. Use a low temperature, not more than 110 to 120 degrees Fahrenheit. Turn the cones over now and then, and eventually most of them will open.

Baking cones sounds weird, I know, but it's really very natural. Many species of pine hold on to their seeds until after a forest fire. The fire clears the ground, lays a bed of ashes, and the heat of the fire liberates the seed, which falls into this hospitable bed. In the absence of a forest fire, your oven will provide the heat necessary to liberate the seeds. A nice dividend of baking cones is the delicious smell of pine, redwood, or whatever that will fill your kitchen.

By the way, if your oven doesn't go down to 110 or 120 degrees, you can try a trick I learned from an amateur yogurt maker. Turn the oven off completely and run a light bulb on an extension cord into the oven compartment. The cord will prop the door open slightly to provide good ventilation, and the bulb will supply enough heat to pop the cones.

DRYING, STORING, AND TREATING WILD SEEDS

If you are going to plant your seeds immediately after collecting them, you might want to skim through this section. But at least glance at the paragraphs headed *Treatment* so that you'll know what to do if your seeds stubbornly refuse to germinate.

Drying   If you're going to store the seeds for more than a month or two, you must keep them dry. Spread them out in shallow layers in a dry, well-ventilated room and turn them over every day or so. Remove any seeds that look moldy.

Or, better yet, dry the seeds in the sun. The sun will kill insect eggs and fungi.

Storing   In most cases, seeds can be kept in paper bags or manila envelopes (not plastic bags!) in a dry, relatively cool place.

If you're going to keep the seeds longer than a few months, the *ideal* way to do it for most Temperate Zone (nontropical) seeds is to pack them in airtight containers and store them in a refrigerator or some other cool place. Do not pack them in airtight containers, however, unless you are absolutely certain that they have been dried properly. Otherwise they'll rot.

The best general advice I can give about drying and storage is this: be careful, but not overscrupulous. Most seeds are remarkably hardy — evolution has made certain of that — and they'll usually survive whatever minor ineptitudes you inflict upon them.

Sequoia
cone and seed

Treatment Most seeds germinate without any particular hassle. But quite often you'll come upon a plant with an especially stubborn seed. Paradoxically, in many plants the *failure* to germinate easily is an important survival trait. Take the case of a plant that flowers in August and whose seed ripens in September. The seed falls to the ground, and if it were to germinate immediately, the tender young shoot would be killed by the winter frost. The seed must stay dormant until the conditions are right. Dormancy consists of a thick coat in some seeds, or certain chemical *germination blocks* in others. Through time the seed undergoes experiences that wear away the thick coat or break down the germination blocks. The seed may sit in acid soil for many months; it may be subjected to the cold damp of winter, perhaps the heat of summer, and often periods of alternate freezing and thawing. In other words, there are lengthy, risky processes in nature that break down the seed's dormancy and induce germination at (hopefully) the right time — and it is such processes that we are imitating with our various treatments.

At the end of this chapter I mention several books that give lists of wild seeds and suggest various treatments to help get them to germinate. Depending on the species, the books might recommend *hot water, scarification, sulfuric acid,* or *stratification.* Here is what those terms mean.

HOT WATER Boil about four cups of water to every cup of seeds. When the water boils, turn the heat off, wait a couple of minutes (until the temperature drops to about 180 degrees Fahrenheit), and add the seeds. Let the water return to room temperature and let the seeds soak at room temperature for another few hours.

SCARIFICATION Cut into the seed coat with a knife or prick it with a pin. For small seeds, rub between two sheets of sandpaper. The idea is to break through the seed's thick coat so that moisture can enter.

SULFURIC ACID Soak the seed in concentrated sulfuric acid for the recommended number of hours (usually two to four). Use enough acid to cover the seeds. Concentrated sulfuric acid is corrosive, dangerous stuff, so be supercareful. When you're through soaking the seed, pour the acid off and save it for the next batch. Wash the seeds thoroughly in lots and lots of cold water. (If, by the way, you can't get any concentrated sulfuric acid, you might try soaking the seeds in battery acid. Battery acid is diluted sulfuric acid, so you'll have to soak the seed for longer than the books recommend.)

STRATIFICATION   Put the seeds into a moist (not wet!) medium. Sand is best. Pack the seeds sparsely so that each seed comes in contact with some of the medium and refrigerate for the recommended period of time (usually about three months). The best container for the seeds (although not necessarily for your refrigerator) is a wooden box covered by waxed paper to prevent rapid drying. I've heard of people using jars but I've never gotten good results this way. If you live in a cold climate, you would do very well to put the moist sand into flats or boxes and leave them outside all winter in a sheltered place. In either case, check them every so often to make sure that the medium is still damp and water them every month or so. *Almost all tree seeds seem to benefit from stratification.* In fact, most nuts, acorns, maple wings, etc., cannot be dried and must be stored in this fashion.

As you try your hand at stratification, sulfuric acid, scarification, or hot water, I hope you'll bear in mind that seed treatment is not an exact science. If at first you don't succeed, experiment! Also remember: after you treat the seeds, they're ready and raring to go. You should plant them at once, so plan accordingly. In other words, don't treat seeds in November if you're not going to plant them until March.

PLANTING

Starting seeds in a nursery   If you want to be very efficient, and if you have lots of time and energy, you can plant seeds in seed beds or flats and raise them artificially before putting them out in the field. This is very good practice for trees, shrubs, and many hardy perennials. There are many ways of doing it, using seed beds, cold frames, etc.; any beginner's guide to gardening will tell you how. But as you follow the gardening instructions, please keep in mind this important difference. Garden plants are often exotic, specially bred plants that are very finicky. They usually need exceptionally fertile soil. Wild plants do not need such fertility. In fact, if you pamper them too much in the nursery, they'll grow lush, wanton vegetation that will make it hard for them to adjust to their ultimate fate in the wild.

Direct planting The seeds of annuals and biennials, as well as those perennials you don't want to fuss with, can be planted directly where you want them to grow. Choose locations that approximate the places where you found the parent plants. Clear a small patch of ground, spade it, rake it free of lumps, compress it somewhat, press in the seed, and cover lightly with soil (sifted soil or sand, if possible). Don't bury the seed too deeply, and *don't plant the seeds too thickly*.

Finally, if you've got it in you, talk to the seeds as you plant them. When I would give our volunteers packages of seeds to plant, I'd suggest that they talk to their seeds. Some kids looked at me oddly, but I've heard many intelligent and touching speeches. I frankly was never sure whether it was the seeds or the kids who derived more benefit from these speeches.

Waiting This is the roughest part. After you plant your seeds, you wait to see what comes up. Did you pick the pods too early? Leave them out in the sun too long? Store them too wet? Plant them too late? You wait and you wait, and it's really quite exciting.

Sometimes everyone around you is growing poppies by the zillion. You take all sorts of care and precautions, but you can't seem to get a single poppy to pop. Do poppies hate you? Examine that possibility. If you feel all's well between you and the poppies, then all I can say is, Wait!

Some seeds stay in the ground a year, or even two years, before germinating. Others (biennials) come up one year in vegetative form, the second year in flowering form. Some, like trilliums, are virtually unrecognizable for several years. So wait! It's happened time and again that seeds we planted years before — and had long given up on — suddenly one spring weekend explode all over the place.

It's worth waiting for.

READING

Wildflower propagation and gardening with native plants are subjects that seem to encourage locally written, privately published books that are often quite good. Be sure to check with your local library.

Also, don't hesitate to ask for information from local botanic gardens and garden clubs. People there are helpful, knowledgeable, and usually delighted to see a new face.

Here are some books on seeding that I have found especially helpful.

Collecting and Handling Seeds of Wild Plants, by N. T. Mirov and Charles J. Kraebel. Forestry Publication No. 5. Washington, D.C.: Civilian Conservation Corps, 1939.

Available in forestry and agriculture libraries, its forty-one pages can easily be photostated. This is an extremely valuable book that covers the collecting, handling, treating, and planting of wild seeds, as well as other means of propagation, instructions for setting up a nursery, and methods of cultivation. This book was written primarily for CCC people in the field, and it's especially strong for California and the West.

Handbook of Wild Flower Cultivation, by Kathryn S. Taylor and Stephen F. Hamblin. New York: Macmillan, 1963.

This is an excellent book with lots of general, easy-to-digest information on how to propagate wildflowers. There are some very good hints for most common flowers. The book seems especially oriented toward New England.

Growing Woodland Plants, by Clarence and Eleanor G. Birdseye. New York: Oxford University Press, 1951. Reissued in a Dover paperback edition, 1972.

This book covers eastern wildflowers along with exceptionally good sections on growing native ferns and orchids. Beside seeds, it deals with other methods of propagation like root division, tubers, and cuttings. Its most valuable feature, however, is the care it takes to describe the soil conditions preferred by each plant.

Seed Propagation of Native California Plants, by Dara Emery. Leaflets of the Santa Barbara Botanic Garden, vol. 1, no. 10. Santa Barbara, Calif., 1964.

Very accurate, but unfortunately very local. This was my seed bible at Redwood Park, since it tells how to handle the seeds of just about every California native plant.

❧7 *Plant Midwifery*

One of my favorite projects was to get together with a group of kids and spend the morning running around in the woods, helping the plants to reproduce. In case you've forgotten your own childhood obsessions, nothing intrigues a kid quite like the mysteries of reproduction.

"C'mon," I'd yell. "Let's go out and see how the plants make babies."

Seeds, of course, are the major means of reproduction, but there are other ways — ancient ways by which plants have reproduced themselves long before they evolved the newfangled techniques of pollen, flowers, and seeds.

We notice, for example, how some trees send suckers from beneath the ground — and this is our inspiration for collecting root cuttings.

We discover a tree that has fallen over, and wherever its branches touch the ground, new trees are springing up. This is our inspiration for layering and stem cuttings.

Such expeditions into the woods to study natural reproduction are necessary if we are to relate honestly to the plant kingdom when we take cuttings. They help remind us that vegetative reproduction is not some sort of gimmick recently invented by gardeners but a natural process of plants. Our role in propagation is to watch

104

Crown suckering

closely, understand what we can about the needs of plants, and quietly help the plants fulfill those needs. We are not plant producers but plant midwives.

How do these projects turn out? I have had many successes and some failures. I could, I suppose, moan and groan about the failures, but I'd much rather celebrate the successes. Accepting failures without feeling bad is hard for many people — especially kids. They think they must be 90 per cent right to get an A, and unless they get an A they are not quite good enough. Nothing in nature is 90 per cent right. A redwood tree produces millions of seeds to get one or two trees that will survive. Why should we make standards for ourselves that are so different from anything else in nature?

Once you get into it, I think you'll find that collecting cuttings is fun. In fact, it can be addictive. When the primitive regions of the mind discover that you can lop off a branch, stick it in the ground, and end up with a tree — watch out! I know people who, when the urge is upon them, run around with knives and baggies in their pockets, eyeing every bush and flower, as turned on as kleptomaniacs in Woolworth's.

If the cutting mania takes you over, go to it! You'll probably try to reproduce every plant in sight, and in a few months you'll learn more than this or any other book will ever teach you.

Wild forests have always been good places for me to get my head straightened out. Acts of growth and reproduction are happening all the time, freely and openly. The complex, awesome dance of creation is always going on. When I'd enter the woods with my crews of little earth mothers and earth fathers to take cuttings, it was with the spirit that we were not the owners or even the guardians of the forest, but that we, too, had come to join the dance.

ROOT CUTTINGS

It's a common sight to find a fallen tree with its network of roots clawing the air. Or to come upon a road or stream whose eroding banks expose the naked roots of the trees. Or in the course of transplanting a tree to leave behind dozens of severed roots. These are sad sights, but you can turn them into an advantage. Cut off some of these roots, handle them properly, and with little effort you can create a whole forest of trees.

I know that poplar, black locust, sumac, sassafras, most shrubs of the poppy family, and many conifers will gladly reproduce this way, and I suspect that many other trees and bushes will too.

Collecting root cuttings   You can collect your cuttings almost anytime at all. But the best time is from late fall through the winter, when the roots are relatively inactive — and when perhaps you are relatively inactive too.

The ideal root cutting to look for is one that is young but not too skinny. A skinny root has too little food reserve to do all the hard work necessary to turn itself into a complete plant. A good cutting is at least as thick as a pencil and from about two to four inches long.

As you cut through the roots (lopping shears or hand snips are fine), be clean and decisive. No ripped or jagged ends, please. The perfect root cutting will be moist, firm, of good color, and free of rot. It may have a light green layer just under the skin. If it has hair roots clinging to it, do your best to keep these roots moist — although this will not always be possible.

Polarity   As you are cutting, you'll have to record which end of the root is nearest the crown of the tree. This end will eventually produce the stem, while the lower end will send out rootlets. If you plant a root cutting upside down, it will get confused and die. The conventional way of recording which end is up is to cut straight across at the crown end and to slant the cut on the root-tip end of the cutting.

Wintering   Root cuttings do best when planted in warming soil. Thus if you collect them during the fall or winter, you should store them until the following spring.

First tie them into bundles like stalks of asparagus. Then bury them in moist sand or sawdust. Keep them in a shed, a garage, or in a cellar. The ideal storage temperature is 40 degrees Fahrenheit, but as long as they aren't subjected to prolonged freezing or prolonged warmth, I wouldn't be too fussy. Keep them covered and leave them alone, except to check every so often that they are still moist.

Darkness Roots, the gentle, sensual, feet-mouths of plants, hate light. That's why they grow down rather than up. Scientifically, I don't know whether it makes any difference, but from the moment I collect them until the moment I return them to the ground, I respect their fear of light and do my best to keep them in darkness.

Getting ready In early spring, as soon as the ground has thawed, you and your root cuttings will be ready for planting. Pull them out of the sand or sawdust, untie the bundles, and lay them out on a damp newspaper in a dimly lit room to examine them. If nothing has happened, just put them aside for planting. Many of the cuttings, however, will probably have a crown of buds around the top and perhaps a dangle of rootlets at the bottom. Decide which is the strongest bud and (as much as it may pain you) flick the other buds off. Leave the rootlets intact, and be sure to keep them moist until planting time. That's important: don't let those fragile rootlets dry out for even a minute.

Planting If you want to be extra sure your cuttings will survive, you can plant them in containers or in good garden soil. Water them, weed them, and tend them for a year until they're ready for transplanting in the wild. Any good gardening book will tell you how to do it: follow the instructions for planting bulbs.

Or you can do what I do, which is to put the root cuttings directly into the field at an auspicious time of the year and let them fend for themselves. Think of your cuttings as temporarily decapitated bare-root seedlings (which is exactly what they are) and do what I

straight cut on crown end

slanted cut on root end

1"

describe on pages 127-34. Remember especially to plant the cutting right side up. Leave the top of the cutting about one inch *below* the surface of the soil.

Afterwards Your cuttings will certainly appreciate some attention during their first six months. Deep watering, a gift of mulch once the stem has broken ground, and protection from nibblers are not necessary, but they will make life easier for your cuttings. They'll also give you an excuse to partake of the struggle of the bud as it bursts through the soil and reaches hungrily toward the sky.

LAYERING

We've all seen how ivy, blackberry canes, and the branches of many shrubs touch the ground and take root. This, basically, is what layering is all about. You bend a branch and bury a portion of it. The buried part grows roots while the tip of the branch takes off like a kid running away from home. You can either leave the new plant in place or return several months later to dig it up and transplant it somewhere else.

Layering is a natural, effortless, and almost foolproof way of reproducing many plants. You don't need any skills, special tools, or a saintly commitment to water, weed, shade, or cultivate. In fact, you can lay this book down, grab a trowel, walk out into your back yard, do the whole layering trip, and be back in your easy chair before a cup of tea gets cold. Several months later, your negligible effort will result in a new plant. It's that easy! I hope you'll try it soon.

When to layer Almost any woody plant will layer successfully, as long as its branches are flexible enough to be bent into the ground. Probably the best time to layer is in the early spring, before the explosion of buds. Use the dormant branches of the previous year's growth. Spring soil is warm and moist, and roots just love warmth and moisture.

If the dormant branches are too stiff and woody to be bent into the ground, it will be best to wait until summer. Use the new growth that is almost mature but still pliable. Summer layering will work well as long as nature or a hose can keep the soil moist enough to promote rooting.

Preparing the branch Take the branch you are going to layer gently in your hand. Before doing another thing, I hope you'll have the courtesy to explain to the branch what is about to happen. Tell the branch you are going to set it half free from the mother tree, and you are confident the branch will work hard to produce its own roots and become independent.

Next, make a sharp bend in the branch about eight to twelve inches from the tip.

If the underside of the bend has cracked from the bending, that's fine. If it hasn't, you should create a wound. You can either cut the underside with a knife (some nursery people cut as much as halfway through the stem), or you can rub off some of the bark that has crumpled from the bending. Wounding seems barbaric, I know, and while it's not 100 per cent necessary, it does serve a purpose. The branch tip will still draw the water and minerals it needs from the tree's roots, since these flow through the inner parts of the branches. But the foods and growth hormones manufactured by the leaves pass through the outermost parts of the stem. Instead of going back to nourish the mother tree, they are interrupted by the wound you've made. They begin to accumulate there, and eventually they stimulate the growth of roots.

Burying the branch Next, use a shovel, a trowel, a mattock, or a pick to dig up the earth and bury the bend of the branch. Use a forked stick, a bent wire, or even a stone to prevent the branch from springing up again. Cover the bend to a depth of three to six inches and stomp the earth over it once or twice lightly.

This will usually be enough, unless the ground is exceptionally clayey. If so, you might wish to dig fairly deep and work in some

compost to lighten the soil and make it more congenial to rooting.

The final step is to put a stake next to the new tip and tie the tip lightly to it to make it grow upright.

The nicest thing about layering is that unless you are up against brutal summer droughts, you don't have to do another blessed thing.

Transplanting   If you want to transplant the new creation, return in about seven to twelve months. Cut the connections with the mother plant and then treat the new plant as you would any other tree or shrub. Dig it up carefully, following the instructions on pages 141-49. Be sure to follow *all* the instructions, especially in regard to pruning the plant back a bit.

Another possibility for a layered plant is mentioned on page 12. You can leave the newly rooted shoots in place. Don't even even bother to sever the connections. That way you can convert a modest bush into a complete tangle. *Garden Beautiful* types won't think much of this accomplishment, but your wildlife will be pleased as pie over the tangled living brush pile you've created.

WILLOW STAKES

One of the first things the early settlers did when they claimed a piece of land was to put up a fence. To make the fence, they'd fell some relatively valueless tree, like a willow, perhaps, and cut it into posts. After driving the posts roughly into the ground with a maul, they'd set the log rails on top of the posts, and there would be a crisp, clean-looking fence — for a couple of months at least.

Now if you've ever dealt much with fences, you know that the major problem is usually decay. But if the fence is made of willow posts, there is another very different sort of problem. After a few months the fence posts begin to sprout. Thick, turgid buds appear and spread up and down the posts. The buds burst into leaf, and soon the fence begins to grow — no longer a fence but a living, vigorous row of willow trees.

Many river trees like willows, cottonwoods, and poplars have this marvelous, persistent ability to sprout. It's an important part of their survival, I suspect. Many of these trees have long, whiplike, or brittle branches that break off in winter and float downstream. The heavier end eventually settles somewhere in the wet mud and sends out roots, and a new tree begins growing.

This remarkable rooting ability, which proved so disconcerting to early fence builders, can be a great boon to us. A willow branch pounded into the ground will grow anywhere — yes, anywhere — as long as there is enough year-round moisture. Willows will root in the most barren and unstable of soils, which makes them the most valuable tree I know of for erosion control. (See page 73.)

Cottonwoods and poplars can also be rooted if you follow the instructions I'm going to give. But in addition to water they need a richer, "river bottom" type of soil if they are to prosper.

When to plant The best time to plant willow cuttings is in the fall or very early spring — when we call the tree *dormant*. Actually, only the leaves are dormant. The roots continue to grow all winter from stored energy, and when the buds burst in the spring, the new leaves will have a healthy system of roots to provide them with moisture and minerals.

There is a way of planting willows when they are in leaf. The danger, of course, is that the leaves will transpire moisture faster than the growing roots can provide it and the tree will dry out. You can prevent this by clipping off all the leaves along the stake except one or two, and by continuing to trim off leaves all summer long. It's a lot of trouble, and it's a bit risky, but if you can plant only during the leafy season, you might give this method a try.

Collecting and preparing willow branches Follow the instructions for pruning on pages 119-23. Any willow will give equally good cuttings, so don't get hung up on species.

After you collect the branches, cut them into convenient lengths for planting. Don't try to chop them up while you're in the middle of a tangle of willows, but drag the branches out to a clear area where you can set up a chopping block and have enough room to work.

The cuttings should be at least eighteen inches long and at least a half-inch thick. Anything this size or bigger — even up to ten or twelve feet long — will grow, but the bigger the cutting, the deeper you will have to plant it, so beware.

One thing that determines the length of the cuttings is the water table. If you're planting on land that is wet year round, you can use shorter lengths. In our part of California, where it gets dry in the summer, I usually have to cut the stakes five feet long or more so that I can pound them deep enough to reach moist soil.

To cut a branch, lay it over a chopping block and use a sharp ax. At the thicker end (the end toward the trunk), make a point. At the narrow end (toward the tip of the branch), make a flat, straight cut.

It is very important to note which is the butt end. If you plant the willow upside down, the sap will flow in the wrong direction and the cutting will die.

Preparing a hole   If the ground is soft and moist, you can just pound the stake into the ground without any preparation.

If the ground is rocky, however, you might strip the bark too badly by pounding, so you must first prepare a hole — much the same idea as countersinking a hole for a screw. For smaller stakes, you can pound a digging bar or even a crowbar into the ground, wiggle it around a bit, pull it out, and insert the cutting. For really big cuttings, you may have to start the hole with a shovel or a post-hole digger (if you've got one), then use the digging bar after you're a foot or so down.

The ground at the bottom of the hole should be moist, wet, or even flooded. If you are planting in winter or spring, remember that the water table is probably much higher than it will be later in the year, so dig deeper than you think is necessary.

Pounding the cutting in   This step is a mind boggler. I would definitely recommend it as therapy to those "nature lovers" who tippy-toe across lawns, who cannot bear to see a tree pruned, and who otherwise insist that plants are very fragile, delicate pieces of creation. You take your carefully shaped cutting, insert its pointed end into your carefully made hole, and just pound the hell out of it. A heavy wooden mallet is the best tool. Or have someone hold a piece of wood on the flat head of the stake while you pound away with a sledgehammer. The idea is to knock the stake deeply into the ground without splitting the top too much. Split stakes grow, but they tend to dry fast, rot, or (if they live very long) develop badly.

The cutting should have at least half its length under ground, and even two-thirds or more of its length can be buried. If you don't plant it deep enough, there will be too much leaf and too little root.

Browsing   Cattle are notorious for browsing young willows. They'll desert a pile of hay, a bed of straw, the shade of an oak tree, or a field of alfalfa and come running whenever they see a young willow. If there are cattle present, you'll have to fence off the planting.

Wildlife browsing should not be too severe, unless you happen to have an overabundance of hungry deer at the end of a long, hard winter. If this is the case, you'd be best off planting bigger, taller, thicker cuttings, which are less tasty and which can withstand browsing somewhat better.

Have faith   The first time I planted willows, I felt unutterably depressed. After a full, hard day's work, I stood there with a group of kids looking at what we had done. It was a weird, desolate scene. Everywhere around us we saw dead-looking sticks pounded into the ground. It reminded me of an empty drive-in theater, or a municipal parking lot with hundreds of parking meters all over the place. We were all very tired, cold, and discouraged. The kids kept asking me if I thought these stakes would grow, and I said, "Of course" — but only because that was what I was expected to say.

Later that spring the kids returned to the area to camp. What they saw, as they told me later, was so exciting that they couldn't fall asleep that night. The "parking meters" were covered with thick, juicy buds just beginning to burst into leaf.

Since then I've found willow cutting to be one of the easiest, surest, and most rewarding of all projects.

HARDWOOD CUTTINGS

Knowing how to take hardwood cuttings is a valuable skill if you are dealing creatively with wild land. Most wild plants are totally unavailable from commercial nurseries. You can often get Japanese, Chinese, or Canadian maples far more easily than you can get the species of maple that is growing (or should be growing) in your own backyard.

Taking hardwood cuttings is one of the easiest ways of reproducing plants. (*Hardwood* means that the wood is fully matured, or hard.) You simply cut off a dormant shoot, store it through the winter, and plant it in the spring. I know this technique works well with dogwood, ninebark, catalpa, hazelnut, honeysuckle, elderberry, and snowberry; and it's valuable for other species too.

I've found this to be an especially good project for school kids, by the way, since it conveniently corresponds to the school year. The kids can take cuttings in November, see them through the winter, plant them in April, and enjoy the results by June. Instead of a final exam and a grade, they'll have dozens of living plants to reward them for their efforts.

Biologically, taking hardwood cuttings is a lot like layering. The difference is that you detach the shoot immediately rather than wait until the roots have formed. Instead of feeding off the parent, the shoot feeds off its own reserves. It's not as safe and sure a method as layering, but it's often more convenient.

How rooting takes place Why is it that a severed shoot produces roots? Some plants, like willows and poplars, have dormant root beginnings all along their stems. They are exceptionally easy to root. Knock them into the ground, and the dormant roots wake up fighting.

Most other plants however, aren't quite so eager. What happens is this. Under the bark is the green, juicy layer of cambium. When a shoot is lopped off, the cambium within the shoot responds to the injury in two ways. First it covers the wound with a callus, or scar tissue. Next, if the moisture and temperature are at all encouraging, the cambium begins to produce roots at the point of injury. It won't produce these roots, however, until it first makes the callus. Keep this sequence in mind, since it explains why we treat hardwood cuttings the way we do.

Taking the cuttings Pruning and trail clearing are excellent sources of hardwood cuttings.

The best time to collect them is in late fall and early winter. This gives the callus a chance to grow before spring.

Choose dormant shoots from the previous season's growth. Crown sprouts or water sprouts are acceptable — as long as they are fairly typical-looking shoots at least as thick as a pencil. Those that will survive best, however, are regular branches and twigs that have been growing in the sun and are therefore generally richer in stored food than shaded branches or fast-growing sprouts.

After you collect the shoots, you should cut them into sections four to twelve inches long with at least two nodes to each section. Make the top cut an inch or less above one node and make the lower cut at a slant just below the bottom node. Make certain that the cuts are clean too: you don't want your cuttings to look as if you had chewed them off the parent tree.

Heels Some plants root better if you take them with a *heel* — a small section of the larger branch to which your cutting is attached. As you can imagine, taking a heel damages the tree from which you are cutting it. So generally I take heels only from branches I've already pruned.

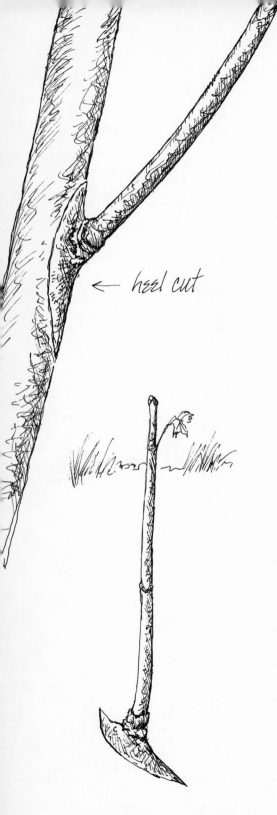

← heel cut

Storing the cuttings You'll now have to store the cuttings until spring. Tie them in bundles, if you want, and bury them in moist (not wet) sand at a cool temperature. The ideal temperature range (if you can do it) is in the low forties — warm enough to form a callus but not quite warm enough to make the leaves sprout.

While your cuttings are cooling their heels, so to speak, you should check them frequently to make sure the sand is still moist and the buds are not swelling. Ordinarily, swelling buds are a cause for great joy — but not now. If you notice this happening, lower the temperature. Otherwise, the shoots will burst into leaf long before the roots have developed enough to support those leaves.

As you check you cuttings, you might also pause to think about what they are going through underneath the moist sand. All its life the cambium has been geared toward producing stem, stem, and more stem. Now it must change. It must totally realign its inner resources and begin to produce roots. Everything looks peaceful and quiet underneath your sand, but the changes are enormous — the sorts of enormous changes that go on inside a cocoon as a caterpillar slowly transforms itself into a butterfly.

Planting Within about six weeks the callus will have formed. Sometimes you can see a scabby white growth. Often it's just a thin, transparent covering.

To be honest, hardwood cuttings look a bit disappointing. "You mean this stick is going to grow?" a kid once asked me.

"It's not a stick," I replied. "It's a magic wand."

You can plant these magic wands as soon as the ground thaws out. Follow the instruction on planting bare-root seedlings, pages 127-34. Insert the cutting deeper into the hole, however, so that the top bud is just peeking out above the surface of the soil. Stomp the ground hard to pack the earth against the stem, wish your cutting *bonne chance*, and you are done. With average luck you can return a few months later and report to your friends a heart-warming tale: Local Cutting Makes Good!

READING

There are dozens, perhaps even hundreds, of books on plant propagation. They all seem to cover the same range of subjects: softwood cuttings, hardwood cuttings, seeds, divisions, grafting, etc. Some books cover the field better than others, but they are all basically sound. If in doubt, choose the one with the most helpful pictures. I mean it.

Among the many books I've looked at, the ones I've found most valuable and practical have been these:

Handbook on Propagation. Handbook No. 24. Revised edition. New York: Brooklyn Botanic Garden Record, 1970.

This short manual (eighty pages) with lots of photos and sketches is the most useful book I know on plant propagation. It tells you most of the important things you should know about propagation and gives many useful hints without drowning you in expert-oriented details. If you can't get it in your local bookstores, you can order it by sending $1.50 to Brooklyn Botanic Garden, 1000 Washington Avenue, Brooklyn, N.Y. 11225.

Plant Propagation: Principles and Practices, by Hudson T. Hartmann and Dale E. Kester. Second edition. New York: Prentice-Hall, 1968.

This is a heavy, 700-page book with lots of information, aimed mostly at professional and university nursery operators. The book is especially valuable, however, because in addition to telling you what to do, it also tells you *why* you are doing it, with long, somewhat technical excursions into plant anatomy, biology, and chemistry. If you want to know more about such things, this book is a good place to turn.

Use of Vegetation for Erosion Control in Mountain Meadows, by C. J. Kraebel and Arthur Pillsbury. Technical Note No. 2. California Forest Experimental Station: U.S. Forest Service, 1933.

I don't know where you'll ever get hold of this one — only bigger forestry libraries are likely to have it — but if you do manage to get a copy, you'll find that it has a thorough section on willow stakes. It tells you not only how to plant them but where and why, with lots of diagrams.

Also, *Growing Woodland Plants* by Clarence and Eleanor Birdseye, mentioned on page 102, has lots of good advice on propagating wild plants by cuttings.

❧8 *Pruning*

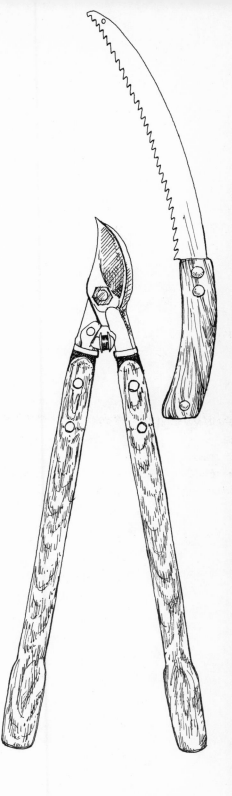

Pruning trees and shrubs is usually a light, easy job. It demands a bit more sensitivity and judgment than most other projects in this book, and it can even be artistic (a bit like sculpting) — but don't let that scare you away. Simply think about what you're doing and, if you are doing it with friends or with kids, avoid anyone who seems overly gung ho. Otherwise, at the end of the day you might find, as I have, a few pathetic, suffering stubs sticking out of the ground, an enormous pile of slash, and some overworked kid with sweat on his brow and an infuriating grin on his face.

While I have run into an occasional mad hacker, most people I've dealt with have been quite the opposite. Most of us are timid snippers, very much afraid of hurting the tree. We seem to think that trees might be like people and that cutting off a branch may be something like cutting off a human arm, leg, or finger. But this is hardly the case. Trees have a certain polymorphous vegetative ability that people lack. If you put a branch into the ground, it will grow roots. If you put roots into the ground, they will grow stems. If you saw off a weak limb, the whole tree will be strengthened and growth will be spurted elsewhere. If you decapitate a pine, it will grow two heads. If you chop a tree right down to the ground, it will often produce dozens of crown sprouts, each striving madly to become a tree.

119

double leaders

dead wood

weak
crotch

There is something very crazy, marvelous, and irresponsible about vegetative growth, especially when you compare it with the careful, conservative growth of humans. I think it's important to understand this difference. The idea that trees are like people leads us to think of trees as an inferior sort of life. Understanding their truly remarkable nature, however, leads us closer to the truth: as wonderful as we are, in many ways we are actually inferior plants.

Why prune a wild tree? I assume, heaven forbid, that you are not going to shape, coif, or barber your wild trees. I am certainly not going to tell you how to shape a bush to look like a dancing elephant, nor will I even tell you how to jazz up an unproductive apple tree you might have in your backyard. A well-pruned tree is perfectly appropriate for a formal garden or a backyard, but it would be as out of place in the wilds as a poodle dressed in a knitted jacket and vinyl booties.

Your attitude toward wild trees should be laissez faire — or at least fairly lazy. Yet there are many times when you will have to prune, such as when you are:

Collecting cuttings;

Building and maintaining trails;

Reducing foliage on newly transplanted trees;

Gathering willow stakes for replanting; or

Repairing the butchering job usually done by highway departments
    and utility companies along their rights of way.

In short, if you *have* to prune, here's how to do it in such a way as to leave the tree strong and attractive.

Tools There are basically only two pruning tools: a wide-toothed saw for bigger branches, and lopping shears for twigs.

Unless you are hopelessly hung up on the "right tool for the right job" mystique, these two tools will serve you very well. Look closely at the fancy arsenal that professional tree people lug around and you will see that all their equipment is just so many variations of these two basic tools.

When to prune   It's probably better to prune a tree when it's dormant, but this is not quite necessary. Just try to avoid the extremes of heat and cold — as much for your sake as for the tree's.

How to begin   If you are repairing damage to a tree, or if you are clearing a trail, you don't have any choice as to which branches you must remove. So go right ahead and do what has to be done.

But if you are reducing foliage on a newly transplanted tree, taking cuttings, or gathering willow stakes, you can prune the tree almost any way you want. Don't just hurl yourself at the tree and begin cutting. Pruning a tree is a very big event — not in your life, perhaps, but certainly in the life of a tree. Plan it out beforehand. Put your tools down as far away from you as possible and sit down near the tree. Look it over, see what it's like, talk to it, listen to it, and get to know it. What is its general shape? How is it balanced? Are its leaves scattered all along the branches or are they growing mainly at the tips? Move around a bit and study the tree from all sides. After a while you will feel thoroughly easy and intimate with the tree; then you can begin the cutting.

I would, as a general policy, put health before beauty, and I would first cut off any branches that are traditionally considered weak: water sprouts, crisscrossed branches, weak crotches, dead wood, double leaders, branches with lots of bulk but little foliage, shaded branches, crown sprouts, and drooping branches.

If the health cut brings you to a good place, then by all means stop there. If not, where you go next will depend largely on your artistic judgment and what your *feel* for the tree is. My own preference is to simplify: that is, keep the general shape and balance but make it more sparse.

Another thing you might do is to *head* the tree. Leave the major branches intact and cut away on the outermost parts of the tree — as if you were trimming a hedge. This is especially valuable for newly transplanted trees, since it keeps the foliage closer to the trunk and reduces the distance that the sap and nutrients must be circulated.

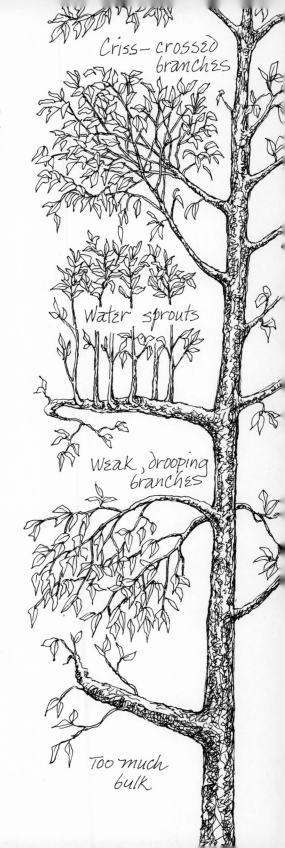

Criss-crossed branches

Water sprouts

Weak, drooping branches

Too much bulk

How to cut a big branch   The most important thing to remember is Don't Leave A Stub! Once you've cut off the foliage, the sap will no longer flow out to the stub. The stub will die, rot, and often carry the rot back into the heart of the tree.

The idea of cutting a big branch is to cut it flush with the trunk, or flush with the major branch to which it is attached. Don't try to do it with a single cut. You'll run into trouble if you do. If you cut from below, the saw will bind as the branch begins to weaken. If you cut from above, the branch may very well tear off a strip of bark as it falls.

The proper way of cutting a big branch is with three cuts, as shown on the opposite page.

The first two cuts will bring down the branch, and the third cut will take off the stub.

Once the branch has been cut off flush against the trunk, inspect the scar to make sure there are no shreds of bark or pockets that might catch water.

Then, if you want, cover the wound with tree paint.

Cutting off twigs   Small twigs (under one-half inch) do not have to be cut off flush with the next major branch, they do not need tree paint, and in fact they do not need any special care at all.

If, however, you want to make the perfect cut, the tree will certainly appreciate your efforts. Use lopping shears to make a slanting cut just above a node. (A node is a place where a leaf or a bud joins the stem.) The cut will heal over quickly and growth will resume at the node.

If you are dealing with kids, by the way, you'll find that they have a tendency to cut off quite sizable branches with the lopping shears, until they have bent the lopping shears out of shape. There's not much you can do about this; it is simply the way kids are made and the way lopping shears are made. Put them together and you get bent lopping shears. I don't have any solution, but I thought I'd warn you anyway.

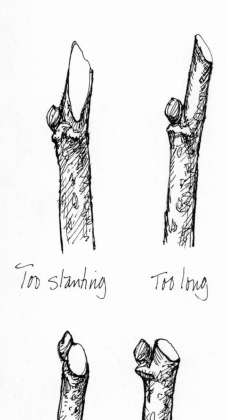

Too slanting   Too long

Too short   Ideal cut

When to stop   I usually like people who get carried away by what they're doing, but not when they're pruning. Knowing when to stop is urgent. Make sure you have a plan, carry out the plan, and then stop. Don't keep snipping away until the tree looks "perfect." If kids are doing the work, or if you're still learning how to prune, the trees you finish will often look ragged, chopped, and impoverished — as if they just stepped out of a barbers' college. Don't feel bad. It will take the tree a few weeks to recover from the insult, heal over its wounds, and begin to make itself beautiful again. Trees want to be beautiful, and they have the power to make themselves beautiful. It's the expression of their vegetative urge. If you've made any aesthetic mistakes, the tree will repair them within a year. Sculptors should be so lucky!

READING

There are lots of good books about pruning, although they deal mostly with ornamental and orchard trees rather than with wild trees.

Sunset Pruning Handbook, by Roy L. Hudson. Menlo Park, Calif.: Sunset Books, 1952. Revised edition, 1972.

The older editions of this book had lots of excellent pictures and drawings. You can learn a lot just by looking at pictures.

The new edition does away with most of the pictures and has instead a very helpful tree-by-tree description of how to prune.

The ideal book would be a combination of these two.

The Pruning Book, by Gustav Wittrock. Emmaus, Pa.: Rodale Press, 1971.

This is a competent book that covers the subject very well. It's not too exciting, but maybe pruning isn't supposed to be. The fact that it's published by Rodale Press will undoubtedly recommend it to organic gardeners.

The Pruning Manual, by Everett Christopher. New York: Macmillan, 1954.

This is a reworking of *The Pruning Book*, written by Dr. Liberty Hyde Bailey in 1898. It deals almost exclusively with fruit trees and handles the subject well. It also has a fairly good section on tree biology as it relates to pruning — very valuable if you're the sort of person who asks why as well as how. You can find it in most libraries, and, since it's still in print, it can be ordered by any bookstore.

9 Planting Trees

The challenge of tree planting, as I see it, is not to find volunteers who will help you out. That's easy. Tree planting is by far the most popular of all conservation projects. Nor is it especially challenging to plant a tree in such a way that it will grow. That's also easy. The real challenge is to overcome the weighty, pious expectations that surround tree planting and to turn it into a fresh, sensual, fun-loving act.

One way I found to make it fun, especially for kids, was to make it easy. I always gave kids fewer trees than I felt they could handle. At the end of the day they were still full of energy and perhaps disappointed that I didn't have any more trees to give them. It was a far better way to end the day than with a group of cranky, blistered kids who wished they had never come.

The other thing I did was to avoid every possible cliché. I think we've all had it up to here with tree planting sentimentality. Never once in dealing with kids did I dangle before them the image of grateful unborn generations thanking us for our selfless act. Instead, I would assemble the kids and give them what must have been the craziest speech they ever heard. "Be quiet, please. You've got to be quiet. It's lunchtime. The trees are eating. Sh-sh-sh! The trees in the forest are always eating. It's always lunchtime. No wonder the trees are so fat. Just look at the bellies on them. They eat all day long, all night long, every day of the year. Eat, eat, eat.

Millions of mouths, always eating. No wonder trees don't move and run around and jump. They don't have time. All they have time to do is eat. The earth is like a huge banquet table, and all their lives they sit at the table, eating, eating, growing, growing, swelling, swelling."

I would then talk about how the trees were breathing, breathing, breathing, always breathing, millions of noses breathing in and out. I would talk about drinking, digesting, circulating, and communicating. "That bee over there is bringing a sackful of genetic messages from one tree to another."

With my fingers I would explain how trees grow — how they grope and strain at the tips of their branches and at the wiggly ends of their roots. We'd look at trees and notice how the branches writhe and strain and how the leaves spread out submissively before the sun.

The purpose of this sort of presentation was not so much to teach kids facts but rather to shake them up, to make them see, perhaps to impress upon them the stark, wonderful, awesome fact that trees are *alive.*

Sometimes, when I sensed that the kids were really with me, I would suggest that we imitate the trees. We would become a forest. Our legs would be roots, our torsos trunks, our arms branches. We'd adopt various positions, experimenting around, until each of us had achieved a sense of his or her own treedom: some were slender and graceful as willows, some squat and gnarled like mountain junipers, some straight and ambitious like redwoods, some muscular like beeches, some loose and expansive like elms. When we had found our tree personalities, we'd hold still. Like trees, we did not move; yet we were very much alive, aware of our environment and in strangely intimate contact with each other. "Think tree thoughts," I'd suggest. Once we began to think tree thoughts, and perhaps even receive tree thoughts from the forest around us, I knew that we were ready to do the job of planting trees.

But is it really necessary to go through an awareness struggle just to plant a tree? Of course not — otherwise damn few trees would ever get planted. (In fact, I've worked on professional tree-planting crews where I've seen whole forests planted by men whose awareness hardly extended beyond their next pay check.) If you put a tree into the ground with any amount of reasonable care, it will grow. That's part of the aliveness of the tree: it has a will to grow. As tree planters we are merely the servants — and, at our best, the admirers — of that will.

BARE ROOT SEEDLINGS

A bare-root seedling is a skinny creature that has been plucked naked out of the earth. You cannot go to a nursery and buy one bare-root seedling any more than you can go to a supermarket for a single toothpick. You buy them in large numbers (the minimum order from a California state nursery is 500). They are easy to plant, easy to come by, and they are cheap. State and industrial nurseries will be glad to sell them to you. The average price is under thirty dollars per thousand seedlings. They are the best way I know to reforest big areas quickly and cheaply.

bare root
seedling

To plant or not to plant   Despite the many virtues of trees, making a forest is not always a good idea. Take, for example, the case of Grass Valley, a piece of land near Oakland that consists mostly of rolling California meadows. It is a rich, varied environment of snakes, gophers, mice, grasses, and wildflowers — a sort of green, magical place over which vultures hang and hawks glide. There are wooded canyons nearby that shelter deer who browse the meadows at dawn and foxes who stalk mice at night. Perhaps one day it will again support eagles.

Each spring, along with the wildflowers, come thousands of Boy Scouts to camp in Grass Valley. And as a "conservation project," they invariably bring along Monterey pines to plant here. I am truly appalled at all the eager, earnest, misguided hard work that goes into planting these trees and the false sense of accomplishment the kids get as they collect their merit badges. To me these orchards of struggling, out-of-place, ecologically senseless trees are nearly as unwelcome as a tract development.

The idea I want to get across is simply this: planting trees can actually be destructive where the trees will succeed a natural, thriving environment. I would never plant trees on meadows, bogs, chaparral, or on valuable wildlife habitats like old orchards, hedgerows, or bramble thickets unless I had some very compelling reason (erosion control or wildlife problems, perhaps).

In the absence of really pressing reasons, I hope you will plant trees only to *re*forest. That is, plant forests only on land which once held forests in the past. There are millions of acres of forest land that have been logged, farmed, grazed, or mined that can urgently use reforestation.

What trees to plant Before ordering your hundreds of bare-root seedlings, have a long, heart-to-heart talk with your land. Here are some questions you might ask it.

What species of trees did you have before you were cleared?

Can your soil support such trees again?

If not, what native trees once grew here which prepared you for the climax forest?

If you don't know how to ask these questions directly of the land, you'll have to ask it of people. Agricultural extension agents, soil conservation agents, and state foresters can give you free advice. Sometimes it's even good advice, if you discount their pro-lumber, pro-agriculture, pro-game-animal prejudices. Don't let them talk you into some Australian exotic that grows ten feet a year, and steer clear of the latest, genetically improved, straight-from-the-lab hybrid "supertree," developed by a university "especially for your type of land." Make sure you insist on native stock whose adaptability and vigor are guaranteed by thousands of years of evolutionary struggle.

Also, try to plant a variety of native trees on your land, although here again you may have to fight expert advice. In the state of Washington, for example, you'll probably be urged to plant solid unbroken rows of Douglas firs — just the way Weyerhaeuser and the U. S. Forest Service are doing it. This sort of monocultural "tree farm" produces good timber, but it is disastrous to wildlife and an insult to the ecological integrity of the land. Plus you run the risk of losing the whole plantation from a single disease, insect attack, or unusual change of weather. The original forests of Washington had western hemlock, spruces, alders, willows, cottonwoods, larches, true firs, and many other minor trees, in addition to Douglas firs. If I lived in that area, I would try to re-create this original variety as closely as possible. I might even give greater weight to the deciduous hardwoods, since they provide a better wildlife habitat than conifers.

There is a trick to planting a variety of trees, however. If you plant them mixed up, one or two species will quickly grow up, over-shadowing and killing the others. If, for example, you were to plant redwoods, live oaks, and buckeyes together, you would end up in thirty years with a sparse redwood forest. To allow the oaks and buckeyes the sun they need, I'd plant them in blocks (or groves) of at least twenty trees.

Spacing How far apart should you plant your trees? Here again, the commercial foresters have got it all figured out. If you space the trees six feet by six feet or perhaps six by eight under good conditions of light and moisture, the trees will grow tall and straight, prune themselves neatly of lower branches, and produce the optimum amounts of cellulose fiber per acre per year. To which my own reaction is, yeccchhh!

You may plant them close together (say, five by five or even four by four) if you feel the trees are puny or the environment is hostile. You should also plant them close together for erosion control where you want a thick mat of roots to form. You can always thin later.

If you expect most of the trees to survive, however, plant them further apart. Ten by ten or even twelve by twelve is good spacing. These trees will keep their lower branches, which make for good wildlife cover, and there will be enough space between the trees for herbaceous plants, berries, and "volunteer" trees.

Another advantage of wide spacing is that trees with low

branches will always be unattractive to loggers — an important survival factor in any forest you plant.

If you decide to adopt the closer commercial spacing, I hope you'll leave clearings here and there for wildlife.

Whatever you decide about spacing, remember that these numbers (eight by eight or six by six) are only rough guides. Please don't plant the trees in straight rows, each tree exactly eight feet or six feet from its nearest neighbors. Imitate nature. Be an anarchist.

Ordering your trees Order from a nearby nursery, one that is within the same genetic "seed zone," and preferably at the same altitude, as your land.

Try to get trees between five inches and nine inches tall. Trees smaller than five inches often get crowded out by weeds — a truly ignoble fate. Trees bigger than nine inches usually have a shaggy, luxuriant root system that makes planting very difficult.

Arrange delivery of the trees as close to the planting day as you dare. If the trees arrive too early, you'll have a storage problem. (Of course, if they arrive too late, you won't have any trees.) If your tree-planting program extends over many weeks or months, you can often have the nursery stagger the deliveries.

Here is a chart to help you decide how many trees you'll need.

Spacing	Trees per Acre
4 x 4	2720
5 x 5	1740
6 x 6	1200
7 x 7	900
6 x 8	900
8 x 8	680
6 x 12	600
8 x 10	540
10 x 10	440
12 x 12	300
14 x 14	220

By the way, you don't have to rent a twenty-foot van to pick up your thousand bare-root seedlings. You don't even need a pickup truck. A shopping cart will more than do the job. You'll be amazed, and probably disappointed, at how little space one thousand tightly wrapped trees take up.

Bare-root seedlings come in bundles, usually fifty trees to each bundle. The roots are packed in moss, excelsior, or some other moist, absorbent material, and they are wrapped in plastic or waxed wrapping paper to keep the moisture in.

If you're going to use them within three or four days, simply keep the trees out of the sun in a cool, moist place, and they'll stay alive and vigorous.

Heeling in   If, however, you are going to keep the trees for longer than a few days, you'll have to "heel them in." Dig a long V-shaped trench in the shade. Make sure the walls of the trench are moist (not puddly wet!). Break open the bundles and arrange the seedlings along the length of the trench, leaning them against one wall of the V. Make sure you keep those roots moist! The roots should all be below ground level. They should not be bunched, turned upwards, or too intimately intertwined. They should just be hanging naturally along the wall of the trench. Also make sure the stems are separated from each other. This is important, because if you pack them too closely together, the little devils begin to generate heat and can actually smolder. Bury the roots, moistening the soil and tamping it in as you go along to remove air pockets. Make sure that the stems are above ground and that none of the foliage has been buried.

The trees are now temporarily planted, and if you keep the soil moist, they will last very well for three weeks or a month.

When and how to plant   The best times to plant bare-root seedlings are in early spring, as soon as the ground thaws, or very early fall, well before the ground freezes. Choose a cloudy, foggy, or rainy day if you can.

heeling in

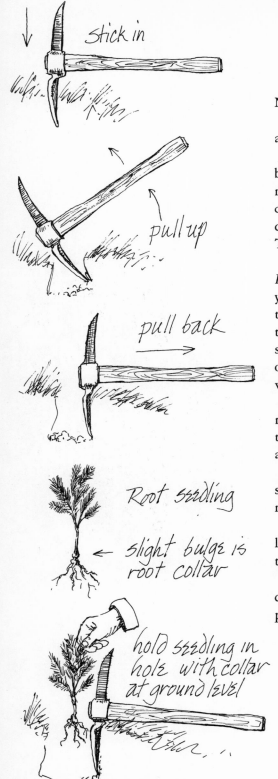

stick in

pull up

pull back

Root seedling

← slight bulge is root collar

hold seedling in hole with collar at ground level

Along the Pacific Coast, where it rains all winter, we've found November through March to be the best months for tree planting.

A strong person can plant 100 trees in about three hours, easily and well.

If you are working with friends or groups of kids, have everyone bring a bucket, pail, or large coffee can. Fill each container with wet moss, wet wood chips, wet soil, anything wet — except water. You don't want water because it will wash off the bits of soil still clinging to the tree roots and leave them more "bare" than necessary. Then hand out the trees.

Now comes the single most important thing to remember: *The Roots Must Be Kept Moist At All Times.* Etch that thought into your consciousness; tattoo it on your arm if need be. Trees feed through their tiniest hair roots, and these must be kept moist if the tree is to stay alive. If you expose the roots to the air on a hot, sunny day for even one minute, the tree will almost certainly die. Even on a rainy day a tree whose roots are exposed for only two minutes will have its survival chances reduced by about 40 per cent.

You and your volunteers now spread out over the field to create a new forest. Your first question will be, "Where is the best place to plant a tree?" One place is probably as good as another, but here are a couple of tricks that might prove mildly beneficial.

Most trees, and especially bare roots, suffer from the hot sun. A spot on the north or east side of a stump, boulder, or other object may give them a slight boost.

A second thing you might keep in mind is moisture. If your tree loves moisture, or the area seems dry, plant in dips and hollows. If the tree loves good drainage, plant on small hillocks.

Each planter will need a planting tool. My favorite is the so-called Western planting tool, or *hodad.* Other good tools are planting bars, mattocks, or narrow shovels.

Here are the actual steps in planting the tree.

1. Scalp the sod if there is any.
2. Plunge the tool deep and straight into the ground.

3. Lift up on the handle.

4. Pull the handle back to open a slit.

5. Peer into the hole you just made. It should be reasonably moist but not filling up with water.

6. Remove one (and only one) tree from the bucket. As you are moving it quickly from bucket to hole, glance at its nakedness. Notice how the roots hang down. When you plant it, you will want the roots to hang down just that way, naturally.

7. Insert the tree into the hole with a wiping motion. Don't cram it in, or the roots will bunch up.

8. Now pull the tree up to its proper height. The root collar should be at soil level and the roots should hang freely.

9. Pull out the tool and let the earth plop and settle around the roots.

10. Next plunge the tool into the ground away from the tree and ram the soil hard against the roots.

11. Withdrawing the tool, stomp hard at the base of the tree to press the soil down around the roots.

12. Pull at one leaf. If the leaf comes off in your hand, the tree is tightly planted. But if you can pull the tree out of the ground by a single leaf, it is too loose — so return to step one and plant it all over again in a different place.

13. Give the tree a second stomp, wish it good luck (aloud, please), and move on.

Mistakes   Here is a checklist of the most common mistakes that cause newly planted trees to die.

DRIED ROOTS   Somewhere in the storing or planting, those tender little hair roots dried out.

"J" ROOTS AND BUNCHED ROOTS   This happens when the hole is too shallow (a common problem if you're dealing with little kids). The roots are either turned up at the ends ("J-rooted") or scrunched up in a tangled ball at the bottom.

WRONG DEPTH   If the tree is planted too deep, with the root collar

pull out hodad and plunge in further back

push forward

heel in

and, one last tug to see that it is firmly planted

buried, the tree may suffocate or suffer rot. If the collar is too exposed, the roots may dry out.

AIR POCKET   This happens when the tree is planted too loosely.

Occasionally you will get a bundle of trees with outrageously lush roots. This is common with two-year-old seedlings that have never been transplanted in the nursery. Remember not to J-root them or bunch the roots into the hole. The ideal solution is to dig a deeper hole, but if this is too demanding, you'd be well off to prune the roots to a more convenient length. Cut them off cleanly with a knife. I don't recommend this, mind you. I just mention root pruning as a lesser evil than J-rooting.

Afterwards   You've planted your land, and as far as I'm concerned, you're through. Mulching? Watering? Deer and rodent control? Fertilizing? If you want to go through the effort, that's fine indeed. Any effort you make will most likely increase your trees' chances of survival. As for me, I feel that bare-root planting is simply a once-over treatment for a large area. Under ideal circumstances, I expect no more than 80 per cent survival, and I overplant accordingly. In rough situations I'm not at all ashamed of 15 per cent survival.

I plant a tree as well as I can, and when I'm done, I turn my back on it. I view myself as a marriage broker: I introduce the trees to the environment in the best circumstances possible, but after the introduction, it's up to them to work out a satisfactory relationship.

CONTAINER-GROWN TREES

Some day, go to your local arboretum and think about what's happening there. You'll probably find loblolly pines from the Caribbean, white pines from New England, ponderosa pines from Colorado, bristlecone pines from California, South American pines, Asian pines — trees from all over the world growing, indeed thriving, right next to each other. For me, it's an inspiration: here I am, fretting, worrying, and chewing my fingernails about whether I can grow redwoods on prime redwood land, and these people are growing trees thousands of miles from their native environments.

Of course, they don't do it the way I described in the previous section — with bare-root seedlings. If the founders of an arboretum had planted their area with a grab bag full of bare-root exotics, no matter how carefully they did it, 90 per cent of the trees they planted would have probably been wiped out within a year. Bare-root seedlings are for reforesting huge areas cheaply and easily, using native stock which stands a good chance of "taking," and which requires no further care to survive.

But let's say that you are planting a more difficult area. Or a smaller area, like a school yard or an empty lot, where you want to be certain that each tree you plant survives. Or perhaps you have a lot of time and energy to devote to preparing the ground, planting the trees carefully, and caring for them later on. In such cases you'd be better off doing what arboretum planters or landscapers do — plant trees that have been grown in containers.

In this section I'm going to explain how to plant a container-grown tree with virtually 100 per cent likelihood of its survival. I'm going to give you a lot of fussy advice about how to prepare the planting area, dig the hole, break the tree loose from its container, what sort of soil to backfill the hole with, how to mulch, how to water, and how to stake and tie the tree so that it won't blow over. I hope you'll view these instructions as you would items on a supermarket shelf: just because they're being offered doesn't mean you've got to take them. If you were to plant a potted tree next to a bare-root seedling and walk off, leaving them both to fend for themselves, the potted tree would be more likely to survive. After all, it still has its root ball intact, a friendly mass of soil to which its roots can cling and draw nourishment thoughout the trauma of "transplant shock." In other words, this is not how you *must* plant every container-grown tree, but the more suggestions you follow, the better off your tree will be.

There's one more thing that I really want you to understand. Too many people get so carried away by the poetry of planting a tree that they ignore the hard-work aspects of it — until halfway through the project, when blisters rise higher, spirits ebb lower, and everyone wishes he or she were home watching TV. There is ground that has to be cleared, deep, wide holes that have to be dug, and perhaps soil, mulch, and buckets of water that have to be hauled to the site. If you have lots of trees to plant, line up as many volunteers as you can. And if you're working with kids, remember: chipping away at a hard piece of ground with a too heavy tool is *no* kid's idea of a good time — even if it is for a good cause.

When and where to plant   You have a lot more latitude with a container-grown tree than with a bare-root seedling. If you're going to take care of it later — that is, mulch and water it regularly — you can plant it almost anywhere and in almost any season (except in the heat of summer and, of course, when the ground is frozen solid).

If, however, you are going to leave it pretty much to its fate after planting, you should follow more closely the instructions for bare-root seedlings in the previous section.

Choosing a tree   Again, you have a much wider choice of species — witness the arboretum — although as a matter of ecological ethics I would still plant only native stock.

One thing to be cautious about, however, is the age of a tree. If you can help it, avoid a tree that's more than three or four years old. Older trees have more difficulty adapting to a new enviroment. There's also a good chance that an older tree may be *root bound.* This happens when the roots grow too big in an enclosed space. Instead of growing outward and downward, they reach the sides of the container and begin to spiral. If you ever discover this condition while you're planting, do not just put the root-bound tree in the ground and hope for the best. It's likely that the roots will never straighten themselves out properly, even in good soil. The tree may look healthy for several years, but as the crisscrossed, intertwined roots grow thicker, they inevitably choke each other and cut off circulation to the whole tree. The tree actually strangles itself. It's gruesome to contemplate. The best you can do for such a tree as you're planting it is to break the ball of earth apart, untangle and comb out the roots by hand (making sure the roots are always moist), and plant the tree as a bare root. Prune the top branches as severely as you dare, give it as much mulching, watering, and attention as you can afford, talk to it incessantly, and you just might save it.

Another thing you can do for a tree is to harden it gradually to its new environment. Don't take a tree from a cool, moist, shady nursery, for example, and plant it on a hot, dry, exposed hillside without preparing the tree for its fate. In such cases you'd be wise to move the tree gradually, giving it a little more sun and exposure each day, and spreading the change out over as many weeks as you can.

Prepare the hole   There's an old saying that claims you'll make out better planting a fifty-cent tree in a ten-dollar hole than planting a ten-dollar tree in a fifty-cent hole. Pay heed!

As you dig the hole, you should segregate the soil. The top layer of sod, if there is one, can be skimmed off. (You might want to keep it as a special treat for your compost heap.) As you begin digging, you should carefully pile the topsoil next to the hole. If you eventually reach rocky subsoil, shovel it away from the planting site.

The hole should be about twice as wide as the diameter of the container and perhaps half again as deep as the depth of the container — unless by some unheard-of good fortune you're digging into good, loamy soil, in which case you can make the hole much smaller.

Now that you've laboriously dug a nice, deep, wide hole, a hole you can be proud of, you should begin to fill it. This is not just make-work, à la U.S. Army. You are going to replace some of that rocky subsoil you may have removed with the best topsoil you have. Use the topsoil you took out of the upper parts of the hole if there's enough, or add your own special concoction of compost and loam. If you don't have any superblend available, do the best you can — perhaps mixing some well-rotted manure or a handful of peat moss with the subsoil. Using whatever you can, build up the hole, moistening and tamping the soil as you fill, until the hole is the same depth as the container.

At the risk of belaboring this point, let me say once more that the soil you put into the bottom of the hole is very important. If it's too sandy or coarse, water will drain too rapidly from the roots. If it's too clayey, the roots will get waterlogged. But if it's just right, the roots will become fat and prosperous, working their way downward and spreading good health throughout the whole tree.

Planting the tree　Once you've prepared the hole, try to get the tree out of its container without crumbling the ball of soil. It helps greatly if you water the tree the day before the planting.

The way I've always done it is to cut the can away from the soil by snipping through the upper rim with wire cutters, running slits down the sides, and pulling the can away. If you want to save the container, you'll have to run a long knife or machete along the inner walls of the can, bang a bit on the bottom and sides to loosen the soil, and carefully slide the soil ball out.

Some nurseries now sell trees in biodegradable containers made of peat moss or tarpaper. In that case, simply plant the tree in the ground, container and all (according to the instructions), and the container will quickly decay. I say "according to the instructions" because in the case of tarpaper I don't believe it, and I very doggedly strip off at least half of the tarpaper or slash it here and there with a jackknife before planting.

Once the tree is out of its container, you should work quickly. Lower the root ball into the hole. The root collar should be level with the surrounding land.

Now fill in the sides of the hole with the best soil you've got. Again, you can use the topsoil you shoveled out of the hole, or more of that superrich compost loam concoction you've brought along for the occasion. No commercial fertilizers, please; you'll burn the roots.

Fill the sides of the holes, pressing, moistening, and tamping the soil in place as you go to eliminate air pockets.

When you get to within an inch or two of the surrounding land, arrange the earth into a bowl with the tree as an island in the middle. The bowl will collect water during future waterings or rainstorms.

Water the tree Add water slowly, letting it sink in before adding more. After a lot of hard work, this long drink should be a leisurely, restful step — both for you and for the tree.

For a long time I resisted adding anything except pure water. Lately, however, I've begun to use vitamin B-1, available in most garden stores under various brand names. I may be experiencing the enthusiasm of a recent convert, but as near as I can see, vitamin B-1 is virtually a "wonder drug," greatly reducing transplant shock and apparently strengthening the entire tree.

Mulch I've been telling you so many things you should do, it's now a pleasure to tell you something you shouldn't do. Do not clear a big circle around the tree.

Perhaps someone once taught you that you must clear a circle to remove competing vegetation that might rob your tree of water and nourishment. But what they forgot to mention is that the circle of bare earth will bake and crack in the summer and freeze solid in the winter.

If you think that competition will be a problem, instead of clearing, lay a thick bed of mulch around the tree. The mulch will also help retain moisture, and it will add organic matter to the soil. Follow the instructions on page 46, remembering especially not to lay the mulch up against the stem lest rot set in.

Screening, staking, and guying I've lost many trees through deer browsing, mouse nibbling, rabbit chomping, cattle grazing, antler rubbing, and people stomping — all because I've never put up any sort of protective screen around the trees. I'll probably lose a lot more trees in my lifetime, too, because I doubt if I ever will put up a screen. It's part laziness, mostly aesthetics.

But if you really want to protect a tree and don't care how it looks, you can knock three or four stakes around the tree and then wrap a piece of screen around the stakes.

As for supporting a tree, you'll probably never run into this problem. With trees under four feet tall it's unnecessary and pretentious. If the tree's over four feet tall, you shouldn't be planting it

Protective screen

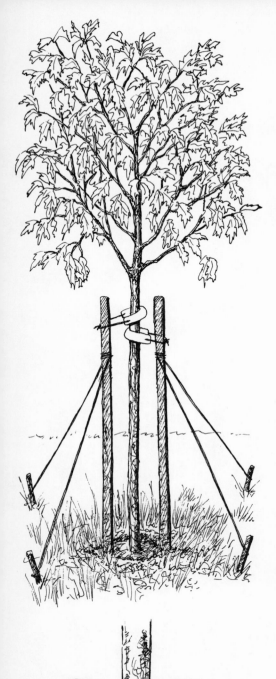

except as an instant landscaping adventure or perhaps as an emergency transplant from an area where it's being threatened.

If the situation arises where you have to support the tree, there are various ways to do it with stakes and guy wires. Check periodically to make sure the supports are taut enough to hold the tree up without cutting into the bark. Also, if you stake a tree, make sure later on that the ground hasn't settled, leaving the tree dangling by its stakes. Remember to remove the stakes within a year, or at most two years.

Transplant shock You plant the tree, you do all the hard work, and at last the tree is in the ground. A week passes by, and you go out to examine your pride and joy. The tree looks just awful. The leaves are droopy, and the tree looks sad and out of place. This is a condition known as transplant shock. Don't worry. You are probably in a greater state of shock than the tree.

If, however, the tree is suffering grotesquely and if it doesn't show any new growth at the tips, you can often help it along with a shot of vitamin B-1 and some courageous pruning. Cut away as much as one-third of the foliage, with an eye toward making the tree more compact (see pages 121-23). This will reduce transpiration (the loss of moisture from the leaves) until the roots have a chance to establish themselves.

Another thing you might do to help the tree along is to erect some sort of temporary screen or other cover to give the tree partial shade from the sun.

Other than vitamin B-1, pruning, and shading, there is nothing else that I'd do for a tree suffering from transplant shock. Least of all would I feel guilty about it. Transplant shock is a natural condition. It is not an indictment of you or your planting methods. So for goodness' sake, don't inflict overdoses of water, fertilizer, and *angst* on a tree that's already got enough problems adjusting to a new environment. Have faith, and as you look the tree over, emit a confident, joyful, optimistic air to which both you and the tree will eventually respond.

TRANSPLANTING WILD TREES

In this section I'll tell you how to dig up wild trees and transplant them. The process is very easy to understand. You simply dig around and under the tree until you've freed the roots with a ball of of soil attached. You wrap the root ball in burlap or canvas to prevent it from falling apart, transport the tree, and replant it somewhere else.

For a very small tree the whole operation may take no more than two people, two shovels, and about twenty minutes. To move a bigger tree you can figure on a whole day's struggle with several of your strongest friends and some equipment.

My advice on this project is: aim small. It's a lot nicer to move a small tree easily and well than to fight all day with a tree bigger than you can handle. In such fights both you and the tree lose.

Where to get wild trees   I am afraid, frankly, that the information on how to transplant wild trees might be misused and that some people will raid the forest for trees. Please don't rip off trees from the wild. Even if you think the tree won't make it in the woods, let it be. The determination isn't yours; survival decisions are best left to evolution, environment, and natural accident.

There are other places where you can legitimately get wild trees. Every time a road is widened, a lot cleared, or a trail built, trees are being destroyed. Find out where the bulldozers are going. Call local contractors. Contact your city, county, or state highway departments. Or monitor your own trail-building activities. Think of the instructions in this section as a way to rescue trees from human "development."

The ideal tree to transplant   If you have a lot of trees to choose from, pick one that is growing in moist soil in a clear, open place. Such a tree will be likely to have a compact, fibrous root structure, which means you can get away with a relatively small root ball. A tree that is growing in bad soil or in heavy competition with other trees will often have a long, spreading, searching root system that makes digging it up very difficult.

Whatever you do, be sure to watch out for trees that are really suckers and crown sprouts — that is, shoots growing out of an old stump. If you start digging one of these, you may find yourself with a six-inch tree and a thirty-foot root system.

Also try to avoid species of trees that send down long taproots. This is especially common with trees and shrubs that grow in dry places.

Some trees that are extremely easy to transplant are maples, buckeyes, horse chestnuts, catalpas, hackberries, hawthorns, ashes, honey locusts, apples, sycamores, poplars, pears, pin oaks, willows, Osage oranges, and elms. I wouldn't avoid trees not listed here, however; moving them is just less easy than easy, but still very possible.

When to move a tree The ideal time to move a deciduous tree is when it's dormant. Of course, only the leaves are really dormant. The roots grow all winter, probably happy that they don't have any leaves overhead to make demands.

Evergreens are (by definition) always in leaf. The best time to move them is in the early fall so that the roots will establish themselves before the soil freezes. Next best time is in early spring, just as soon as the ground thaws.

Of course, if you're out to rescue trees, you may have no choice about the "ideal" time: a bulldozer may have made the decision for you. Even if you're way out of season, go ahead anyway. Just be more careful, dig a bigger ball, water the tree often after you transplant it, prune it back more severely than you might, add vitamin B-1, and you'll most likely make out fine.

Check for disease Before moving a tree, inspect it thoroughly. Examine the leaves, buds, branches, and stem. When you dig, be aware of the soil around the roots. Look the tree over for signs of insects, larvae, galls, slime, rot, fungi, or anything that looks weird. Avoid moving a weakened tree, since it will not transplant well and you don't want to spread any infection or infestation.

Prepare the soil   If the soil is too dry when you move a tree, the ball will crumble. If the soil is too wet, *you* will probably crumble. (You have no idea what "heavy" is all about until you've spent an afternoon struggling with a sodden ball of soil.)

The best way of handling this is to wait until two days after a heavy rain. Or, if the soil is fairly dry, water the tree thoroughly about two days before you are going to move it.

How big a ball   After a while you'll develop a sense of what size ball you'll need for each species and habitat of tree. The rule of thumb is to dig under the drip line (the outermost branches), which is where most of the feeding roots are concentrated. But if, like a disappointed hitchhiker, you distrust your thumb, you can refer to this handy, official table.

Caliber* (inches)	Diameter of Ball (inches)	Weight of Ball (pounds)
less than 1	14	115
1 - 1¼	16	175
1¼ - 1½	18	250
1¼ - 1¾	20	340
1¾ - 2	22	450
2 - 2½	24	600
2½ - 3	28	815
3 - 3½	32	1,400
3½ - 4	36	2,000

*Caliber means the diameter of the tree at a point about six inches above ground level.

This table was written in the 1930s by the American Nurserymen's Association. Everyone who moves trees seems to have a copy of it tucked away somewhere. Everyone calls it The Guide and considers it the final authority. And, to let you in on a trade secret, everyone cheats like crazy on it.

You can get away with a smaller-than-recommended ball if:
The tree is one of those listed previously as an easy-to-move species;

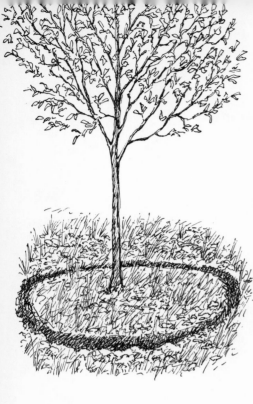

You are moving it from an "ideal" habitat (moist, loamy soil and little competition);

You are moving it at an ideal time;

The tree is an evergreen, most of which have fairly compact root systems;

You are going to take extra good care of the tree after it's transplanted; or

You are willing to take a risk.

In any case, I generally use this guide and the drip line merely as crude estimates, and I proceed on a trial-and-error basis. If as I dig I come upon no roots whatsoever, I assume that all the roots must be within the ball and the ball is probably too big. If, however, I find myself cutting through several major roots without many hair roots, then the hair roots are probably all outside the ball and the ball is too little. A just-right ball is somewhere between these two extremes — lots of fiber and a few middle-sized roots.

Tree roots that have been left behind, by the way, can often be used as root cuttings (see pages 105-8).

Root pruning If you have six months' or a year's warning before moving a tree, there's an interesting trip you can take that will increase the tree's chances and will allow you to reduce the size of the root ball significantly.

Dig a trench around the tree as if you were going to remove it, except that the trench might be closer than recommended. Do not undercut the tree, however. Saw cleanly through the many large roots you will come across. Then fill the trench again with a rich, compost-filled loam, tamping the earth down and moistening it as you fill. Water the tree well and leave it alone. The tree will begin to develop hair roots within the area defined by the trench, thus preparing itself to be moved.

If by some miracle you have two years' warning before moving a tree, you can do the root pruning in two stages, cutting alternate sections of the trench each year.

One word of warning, though: don't root prune any tree in the heat of summer. And don't root prune an evergreen in the dead of winter.

Lifting a small tree   To lift a small tree (one whose diameter is under 1½ inches, say), all you need is a friend and two pointed spades. If the soil is soft enough, you plunge the spades into it and score a deep circle around the tree. If the soil is rocky, you may have to dig a trench. Then, using both spades on opposite ends of the circle or trench, pull on them like levers, slipping them under the tree as you pull and lifting the ball up. Handle the tree gently out of the hole, place the burlap or canvas underneath, pull away the spades, and fasten the cloth tightly around the ball.

Lifting a medium-sized tree   Here's how to lift a medium-sized tree, with a diameter of up to about 2½ inches.

The easiest way to do it is with power equipment like a back hoe, which is not so outrageous as it sounds. When we've rescued trees from road crews and bulldozers, we've often found our "enemies" more than willing to cooperate with us. A skilled back hoe operator can very gently dig out a tree, lift it, and nestle it down onto a piece of burlap which you've laid out on the bed of a truck.

If back hoes are out of your reach, you can, of course, do it by hand — or rather hands, because you'll need at least three or four strong, willing people for the job. First dig a trench at the appropriate distance and to the appropriate depth. Then round off the ball with the back of a shovel, tapering it inward at the bottom. Keep working at the bottom until the ball collapses into the hole.

Usually you will have to wrap the ball before removing it from the hole. Use two shovels on one side (or two shovels and a plank) to tip it as you shove the burlap halfway underneath. Repeat the operation on the opposite side of the ball, pulling the burlap through, wrapping it around the ball, and finally tying it.

Once the burlap is in place, holding the ball together, you can begin to pry the ball out of its hole. Shovels and planks will usually do the trick, especially if you create a slope on one side to move the tree onto.

If you find that you've gotten into something bigger than you can handle easily, you'll have to dig out an inclined plane, slide a platform down the plane and under the ball, and pull it out this way. When building an inclined plane, you might want to water it heavily so that the tree will slide up it on a bed of mud.

Once you get the burlap around the ball, you can temporarily fasten it in place with nails. Then wind some sort of cord around the ball and tie it tight. If you are as inept as I am at knot tying, use several smaller lengths of rope so that if one section happens to slip loose the whole damned thing won't unravel.

Instead of burlap, it is also very common to build a box for the roots out of planks and plywood. If you do so, make certain that the corners of the box will hold together under the tugging and straining necessary to get the box out of the hole. I usually use corner braces. Also, if the plywood is thinner than three-quarter inch, reinforce it wherever you see it beginning to bulge out of shape.

The advantage of wooden boxes is that they hold the roots and soil together much more safely than burlap — especially for larger trees. The disadvantage is that boxes are generally much heavier than balls, and to get one out of its hole you will almost definitely have to build an inclined plane and perhaps use a winch.

Moving a big tree If you have to, you can move a tree as tall as twelve feet with a trunk diameter of four inches or more.

You'll need plenty of equipment — ropes, winches, flat-bed trucks, pulleys, chains, etc. A back hoe is a real blessing. If you have enough equipment, moving a big tree can be an elegant piece of engineering. If you don't, it can be a frustrating ordeal.

Because of all the equipment you'll need, instructions for big-tree moving are out of the range of this particular book. If you want to try it anyway, hunt up a copy of *Transplanting Trees and Other Woody Plants*, mentioned at the end of this chapter. It has lots of nice diagrams and valuable information.

Big-tree moving is very exciting, and it's the sort of monumental project that turns some kids (and adults) on. For several days after you've moved a big tree you feel like a Pharaoh must have felt after he built his pyramid. Better, actually, since you did the labor yourself.

Bare rooting   Up to now I've used a curious fiction that how-to books often toy with — namely, that everything will go smoothly. Now, time out for reality. You gently insert the shovels and planks underneath the root ball, you lift it tenderly to slip the burlap underneath, and the root ball falls apart. Or perhaps the burlap splits or the solid, reinforced box you made shatters under the strain of pulling and tugging. Instead of a neat, compact ball, you are faced with a tangle of naked roots screaming for attention.

Whatever you do, don't scream back. Quickly wrap the roots in damp burlap to keep them moist. Wrap them separately, one piece of burlap for each major root. If you try to pull all the roots together and stuff them into one piece of burlap, you'll injure them.

Actually, it is quite possible to transplant a tree without any soil ball at all. In fact, many — maybe even most — nursery people do not even bother with heavy soil balls if they are moving a deciduous tree while it is dormant. They just dig the tree out of the ground, knocking the soil off the roots and wrapping each root in burlap as they progress. Instead of a root ball they end up with something that looks like an octopus in bandages, but the tree usually survives such indignities.

If you do transplant a tree bare rooted, either by necessity or choice, remember not to let the roots dry out even for one minute, and don't forget your vitamins — vitamin B-1 to be precise. Also, a large bare-rooted tree might be unsteady, so you may have to stake it for a year or two.

Transporting the tree   When you move a tree, make sure it is secured so that it doesn't become damaged or destroyed by knocking around in the back of a truck. It might be wise to tie a lasso around the branches and bring them closer to the trunk.

Also, don't do what I once did — speed proudly along the highway with a load of uncovered trees in an open pickup truck. They all got windburned on the side facing the wind.

And remember to keep the ball moist if you're driving a long way.

Planting Follow the instructions for planting container-grown trees in the previous section, keeping in mind the following suggestions.

Keep the tree out of the ground for as short a time as you can manage. If you're handling a big tree, it would be wise to dig the hole first, then dig up the tree. Try to plant the tree in an environment, altitude, and exposure similar to that from which it came.

It's also important, I feel, to orient the tree properly. Make sure the side of the tree that was originally facing north is facing north again. Otherwise, branches, leaves, and bark that developed in the shade will suddenly be exposed to the sun at a time when the tree can least handle such changes. If you have trouble telling east from west, bring along a compass and tag the tree before you pull it out of the ground.

If the ball is wrapped in burlap, you can loosen the burlap and leave it on. It will deteriorate very fast. Canvas and wood won't however, so they should come off. Just reverse the procedure by which you got them on.

Pruning Pruning a tree is good — it may even be necessary — but you can still hardly bring yourself to do it. It's like going to a dentist.

But when you transplant a wild tree, you have got to prune. The foliage must be reduced by at least one-third to help compensate for the loss of roots. Follow the instructions on pages 121-23, remembering especially to *head* the tree. That is, as you prune make the tree more compact rather than sparser. Prune as drastically as you dare, but remember, whatever you do, don't prune the *leader*. (The leader is the topmost center spire of certain trees.)

Guying, staking, and aftercare Again, follow the instructions in the section about container-grown trees.

One good thing to do, especially for evergreens, is to wash off the the leaves periodically (but not on hot, sunny days). The leaves will absorb some of the moisture, and this helps reduce transpiration.

Transplant shock for wildlings can be severe, and their survival rate is lower than for nursery-grown trees. But don't let this scare you off. If you have planted them well, if you prune them back enough, if you care for them devotedly, most of the trees you transplant will pull through very well.

READING

Most garden books have helpful sections on tree planting. Especially valuable is *Tree Maintenance* by P. P. Pirone, referred to on page 61. Also, for bare-root seedlings and general forest planting, check with your state division of forestry. They often put out very specific guides to the trees and conditions of your area.

Transplanting Trees and Other Woody Plants, by A. Robert Thompson. Tree Preservation Bulletin No. 1. Washington, D.C.: U.S. Department of the Interior, 1940. Revised in 1954.

This is another of the CCC bulletins that was exceptionally well done, with lots of basic and practical information. It has lots of sketches, lots of knowledgeable hints, and a generous section on how to move big trees. You can still get it by sending thirty-five cents to the Superintendent of Documents, U.S. Printing Office, Washington, D.C. 20401. It will answer most of your questions, except the big one: "Why isn't the government doing stuff like this today?"

♨10 *Ponds and Watering Holes*

I've made several small ponds. The one I'm least proud of operated on a float-valve system and looked hauntingly like the tank of a toilet.

But, come to think of it, all the ponds I made were in one way or another artificial. Ponds are not a natural feature of the Oakland Hills, and I've been asked whether building them isn't a "development" — the sort of thing I'm usually bitterly opposed to. My excuse, in case you're interested, is this: Since man has been over-building, overfarming, and generally misusing the soil, he has severely lowered the water table. Streams which once flowed year round now dry up in the summer. By making artificial watering holes and ponds I am merely helping to restore the water conditions to what they were before the arrival of destructive technology.

So much for the excuse. The truth is that I'm hung up on ponds the way some people get hung up on instant cake mixes. You buy the mix in a dull, squarish box, empty out an unpromising powder, add water, and as if by magic you end up with a cake.

That is sort of what it's like building a pond. Begin with some dry land, add water, and stand back! You'll soon get dragonflies, cattails, reeds, rushes, sedges, red-winged blackbirds, frogs, newts, quail, doves, and other assorted wildlife. Damselflies, as pretty as their name, stake out territories. Striders, boatsmen, and whirligig beetles appear from nowhere.

150

Adding water to dry land is the best piece of magic I know. I hope you'll try it soon.

Basically, the way you go about making a small pond is to dig a hole in the ground, seal the hole so that it won't leak, and then fill it with water. This is not as easy as it sounds, since water is a most rebellious substance. If you turn your back on a pile of rocks, you can be pretty sure the rocks won't run off. In fact, rocks will sit dutifully in one place until the next geological age. Not so with water! Water will run downhill, flow over obstacles, flow under obstacles, push things out of its way, sink into the ground, or, as a last resort, it will simply evaporate. Water seems to have a head-strong mind of its own. No wonder kids enjoy it so much.

I found no other project (except chopping down trees) that would turn on a group of kids as completely as building a pond or a watering hole. While there is a lot of hard work to be done, there is also the chance to get muddy, dirty, and wet while having a cover story for Mother: "We were helping the man build a wildlife pond." Nothing ends the day as merrily as a rowdy, no-holds-barred mud and water fight. In fact, one problem I've had with this project is that water fights would often end the productive phase of the day long before I wanted it to be over.

Yet I not only loved to see kids get wet, I encouraged it. In fact, I encouraged them to get acquainted with water on all its levels. To me, water is the most amazing stuff in the world. Try to see it through fresh eyes sometime. Take a glass of water in your hand and pretend it's something brand new — a new element brought back by astronauts from the moon, perhaps. You are a scientist, and you have to describe it for the first time to an eager world. Put your finger in it and describe how it feels to someone who has never felt water. Withdraw your finger and describe how the water closes up over the finger hole and instantly forms a placid surface. Pour it from one container to another and describe the sound it makes. Try to explain the riddle of how it can be both transparent and visible at the same time — how it distorts and reflects, yet how at the same time you can see through it. Watch it drip, spray, flow, turn silvery, cohere, adhere, freeze, boil, and do a thousand amazing tricks. Spend ten minutes playing around with water, and you'll find it to be the most peculiar stuff imaginable. If you happen to be with a group of receptive kids, perhaps you can communicate what a wonderful substance water really is. And if the mood is right, perhaps you can go on to explain to the kids the profound secret that our schools, jobs, governments, and all our institutions conspire to keep hidden from them, a secret we can learn from poets and artists: namely, that when seen freshly, all of the common things around us are crazy, inexplicable, and totally miraculous.

Unlike tree planting, where you have to wait twenty years for a forest to grow, the satisfactions you get from building a pond are immediate. I remember working once with a group of kids to catch water from a spring and pipe it downhill to a sheltered place. It took us three days. When we finally hooked up the last section of pipe and the water came trickling through, a cheer went up so loud and joyful you'd have thought we had just won an Olympic event.

The next day we returned to the watering hole to see if it was filling up and holding water.

"How long do you think it will be before it will attract animals?" they asked me.

"A few months," I answered, as if I knew what I was talking about. I was wrong! I looked, and in the middle of the watering hole was the clear impression of a raccoon track. We stared at it for a long time. You'd have thought we had found the Hope Diamond in our watering hole. The track, of course, is long since gone. But it has left a permanent impression in my mind and the minds of the kids.

WATER SOURCES

To build a pond, you'll need a steady source of water, a site capable of holding the water, and perhaps the plumbing to get the water from its source to the pond site. The hang-up is usually in finding and developing a steady source of water, so let's begin there.

Potholes   The handiest source of water is ground water that is close to the surface. All you have to do is dig until you reach it.

A marsh that remains wet for nine or ten months of the year is your best bet. But before digging into such a marsh, you must make certain that there are no rare creatures whose survival is favored by intermittent water. In our part of California, for example, there are several native frogs that burrow into the mud as the marshes dry out. Once these areas are converted to year-round watering holes, an introduced species of bullfrog moves in and takes over, pushing out the natives. In most areas, making a pothole will be a positive asset to your wildlife, but to make absolutely sure, check with your local state college or some other wildlife authority.

The best way to learn how to make a pothole is from alligators. As the marshes in the Everglades begin to dry out, alligators dig small ponds by thrashing wildly about. During the dry seasons, these alligator ponds support an unbelievable concentration of alligators, turtles, snakes, fish, and birds. In terms of richness, they are the closest things I know in nature to a teeming ghetto.

If your land doesn't have any alligators, you'll have to do the work yourself. Grab a shovel and dig down to the water level. Dig deep enough the first time so that you won't have to disturb the pond again, even when the water table sinks in the heat of summer. As you dig out the pothole, follow the instructions in the section of the chapter entitled CONTAINERS (page 156), especially the instructions in regard to sloping the banks and providing shade, wind shelter, and animal cover.

Another thing you can learn from the alligators, by the way, is to keep your pothole fairly small. If you were to build a big pond, the evaporation would be greatly increased. This could lower the water table for the surrounding area.

On the whole, you should consider your pothole a temporary pond

—one that you may have to dig out every year or two. This adds up to some work, but I think it's worth it. A small pothole will keep your marsh alive and vigorous during the dry spells in late summer. It may be just what you need to turn an intermittent marsh into a real swamp.

Springs   Another natural source of water is a dripping spring, as long as it drips all year round — especially in dry times, which is when you really need the water.

A good place to look for springs or seepages is along the banks of road cuts. If the dripping water is naturally collected somewhere below the spring, I wouldn't do anything further. Nature has made you a perfect watering hole. If the drippings disappear into the ground, however, you can turn to the next section on containers to learn how to seal the ground so that the water will build up into a little watering hole.

Some springs release lots of water, but instead of dripping, the water spreads out over a mossy rock. One thing you can do is to wedge a stick or two in strategic places to lead the water away from the rock and drop it in a place where you can better use it.

The ancients used to think that gods, spirits, and genies lived in springs. Whether this is true or not, springs are certainly temperamental creatures. Hurt a spring's feelings and it will stop running. When this happens, you're left feeling very helpless, and it is easy to believe that deep within the earth a genie is slapping his knee and laughing indecently at your plight.

To avoid insulting a spring, you should do everything quietly and gently, moving as little earth as possible. Also make certain that you do not create any back pressure on the spring. Genies hate back pressure! As long as the water is dripping down into a container, the spring will continue to function. But if you should try to build up a container to the level of the spring outlet, the back pressure will often clog the pores through which the water flows, and the water will seek other outlets.

Seeps Sometimes instead of a flowing or a dripping spring you will find a wet spot on a hillside. You may not see any water, but the ground will be moist and there will be sedges, rushes, or maybe even cattails. You know there is water, but how can you make that water available to wildlife? Get a hand trowel, sit next to the seep, and begin digging near the bottom. Dig a long, narrow hole, like a tunnel, angling it upwards. Do it gently, easily. Every time you take out a scoop of earth, pause for a few moments to see if water will begin to collect and flow down your tunnel. Once the flow begins, stop working and go home for the day. The next day you can return. Has the flow increased? If so, you should stop digging. If the flow has decreased or stopped, however, you'll have to dig some more. Eventually you should get a nice flow, even from an amazingly short tunnel. Push some gravel into the hole — gently, please — to keep it from collapsing. Or, for a more permanent arrangement, insert a length of pipe with holes drilled along the bottom. You're now through. You've turned a seepage into a first-class spring.

Rainfall-catching devices In certain areas of the country you might consider trapping the rainwater as it falls. The idea is based on water-catching systems in dry areas where rainwater flows off a roof into gutters, down a drainpipe, and into a barrel, where it's held for future use. Keeping this model in mind, you might be able to take advantage of natural run-off which you can funnel into a tank or cistern. Or you might make what is known as a "gallinaceous guzzler" to catch dew and condensation. Since I haven't had any experience here, I won't presume to give you advice. The Forest Service handbook mentioned on page 24 has what look like excellent diagrams and suggestions.

Public water supply If you have a water line running through or near your property, you might consider tapping into it. Personally, I feel ambivalent about using public water supplies for a wildlife pond. On the one hand, it seems unnatural, it offends my sense of wild-land etiquette, and I'm dubious about how chlorine and other added chemicals affect the balance of pond life. On the other hand, it's easy to get at, plentiful, dependable, and ever so tempting. I'll leave you to hassle out the pros and cons. I've used municipal water for a pond, and despite my doubts about the source, I was very happy with the result.

If you do decide to use the public water supply, you'll have to deal with the chlorine in it. Chlorine evaporates in about twenty-four hours, so that the day after you fill your pond there will be virtually no chlorine present. To minimize later damage, you should keep the pond level high by adding new water gradually and often. That way the incoming chlorinated water will never be more than a small percentage of the total standing water. If, instead, you let the pond level go way down and then turn on the tap full force, the sudden inflow of chlorine will wipe out a lot of the algae and microlife upon which a healthy pond environment depends.

CONTAINERS

Once you are sure of your source of water, you can begin developing a site for your pond.

Dig in or build up There are two different ideas about how to make a pond. There are those who simply make holes in the ground, and there are those who build walls or dams. I've tried both ways, and unless I lose my sanity completely, I don't think I'll ever again build a pond on the dam principle. Whenever I've done it, I've spent dozens of boring days getting the walls properly compacted and graded, building elaborate spillways, and contending with leakages and cave-ins. I had to keep the walls well covered with sod while stopping the growth of trees whose roots might open up ways for the water to escape. Every time I saw a crayfish, a gopher, a mole, a badger, or any burrowing animal, instead of rejoicing, I'd fret. I even became suspicious of earthworms. An elevated wall that contains water is totally unnatural, and everything in nature will work day and night to destroy it: rain, wind, trees, frost, and animals. Build a dam or a wall and you'll be fighting your land, which is exactly the attitude you want to get away from. So take my advice: avoid berms, retaining walls, dams, and other above-ground structures. Instead dig yourself a humble hole. You'll be a lot happier.

Where to dig a hole   The best place to dig a hole is on level ground that holds water naturally. Clay soil is usually fairly watertight. To find out how well your ground holds water, dig a test hole, fill it with water, and watch what happens. If the soil is exceptionally dry, you may have to refill it two or three times. If after the second or third refill the water stays around for several hours, you're in good shape and can build a pond. As time goes on, the pores in such soil will probably seal themselves, and the site will be relatively impermeable.

If however, the soil is exceptionally well-drained everywhere you look, don't give up. Instructions on how to stop leaks follow later in this chapter.

If you have a wide choice of locations, you might get fussy and look for a place with shade and shelter from the wind. Shade keeps the water cool, reduces evaporation, and discourages the rapid proliferation of algae that takes place in the full sun. Shelter from the wind is desirable since ruffled water evaporates much faster than calm water. If you cannot find a ready-made, shady, sheltered spot, do your best to create one by strategic planting of trees.

Digging a hole   What can I say about digging a hole? It's simply hard work. So get in there with a pick and shovel, get the job done, and promise yourself (and your helpers) a nice reward when you're through. Wheelbarrows are handy, since you'll want to move the loose soil far enough from the pond so that the first rains won't wash it back in. As you dig, slope the sides of the pond gently so that any small animals that fall in can scramble back out again.

How big? That depends, of course, on your flow of water. If you are dealing with a drip, drip, drip, a small "guzzler-type" watering hole of less than one square foot would be appropriate. If you have unlimited water, you can dig as deep and as wide as your muscles and willpower will let you.

If you are planning to establish a pond environment with a few fish and some water plants, you should make certain that parts of the pond are at least two feet deep. This will help keep the pond at a fairly constant temperature and will give fish a cool place to escape to. Totally shallow water heats up during the day, gets cold at night, and freezes solid for a good part of the winter.

Erosion Whenever you are dealing with water, you should think about possible erosion. Standing water won't cause any problems, but moving water definitely will. So study carefully the places where the water enters your pond and where it might overflow.

The water entering the pond can usually be taken care of without much difficulty. Provide a rock for the water to fall over, and this will break the force.

The most serious erosion problems occur at the overflow point. Find the lowest point along the banks of your pond, either by eyeballing it (which can be misleading) or by using a line level.

If you expect overflow to be a rare event, simply make certain that the lowest point is well vegetated, preferably with a thick sod.

If you expect frequent overflow, you should make sure that whatever route the water takes is solidly protected. A bed of rocks (riprap) is very good. It will slow the water down and give it a chance to sink into the ground.

Whatever you do, don't let a gully get started at the pond's outlet. If you see one forming, rush (don't walk) to the chapter on erosion control (pages 63-89) and deal with the problem immediately. Not only will the gully cause a lot of damage to the area surrounding the pond, but it will eventually undermine your pond and drain it.

Leaks If your pond begins to leak, or if you started with well-drained land that won't hold water very well, there are several things you can do:

MUCKING   Very often you can stop a ground leak by mucking up the water. Simply stand in the middle of your pond with a pick, a mattock, or a hoe and make a big, muddy mess. An Irish setter, if you can borrow one, will do the job beautifully. As I discuss in the chapter on erosion control, dirty water clogs the pores in the ground and penetrates only one-tenth as fast as clean water. A couple of episodes of mucking are usually enough to clog up most leaks.

COMPACTING   This is hard work, but it is a natural way of stopping leaks. Let the water drain out of the pond and plow up the bottom. Plow it as deep as you can — at least one foot. A rototiller is good if you can get one; otherwise, use mattocks and picks. Rake out all the rocks, roots, and other vegetation you find. Then compact the bottom.

The way this was done in "the good old days" was with livestock. A herd of cattle, oxen, or horses was driven back and forth until the pond bottom was as hard as pavement. If you can't borrow a herd of animals for the occasion, perhaps you can hold a square dance or convince your local football team to work out on your pond bottom. If all such efforts prove fruitless, do a lot of stomping and pounding with the butt end of a four-by-four. As you can imagine, compacting is practical only for small areas.

ADDING CLAY   The usual reason for leaks is sandy or rock soil. You can usually correct this by adding clay. Mix the clay with the soil to as great a depth as possible and then compact it as well as you can. You can usually get ordinary clay locally, by digging it out of the ground.

A special kind of clay that is used for sealing ponds is bentonite. Bentonite is also used as a base for finger paints, and you can often find it sold in arts-and-crafts supply houses. It is a very fine, powdery clay with a microscopic structure that looks like an accordion. Add water, and each accordionlike particle expands many times its size. If you mix it in with the soil at the bottom of the pond, it will expand until it seals the pores. Bentonite is almost foolproof, but like other "foolproof" things, it has a serious drawback. It has a slimy, creamy texture — like finger paint. So if you use it, dig it in very deep and then dress the surface with dirt or rocks to keep all that slimy stuff underground.

LAST RESORT   If you've tried everything and the pond still keeps leaking, I guess all that's left is modern technology. Concrete, fiberglass, and various plastics should work, but since I've never been that desperate, I don't have much to say about them.

PLUMBING

If the water drips directly from the source into the hole, you can skip this section. Skip it with joy.

If you have to move water from one place to another, however, you'll need plumbing. Even if you don't know a street el from a union, don't worry. Plumbing a pond is a whole lot easier than plumbing a house. In fact, house plumbing has given the whole trade a terrible reputation. It usually involves crawling around on your belly in wet, smelly places, handling clumsy, oversized tools in cramped quarters, struggling with corroded connections of galvanized or even cast-iron pipe, and doing everything according to a complicated system of "codes."

Plumbing a pond, on the other hand, is much easier. The only hard part is burying the pipeline, which amounts simply to ditch digging. The rest is easy if you follow these three hints.

1. Use plastic pipe. Steer clear of galvanized pipe unless you have all the right dies, taps, wrenches, vises, reamers, and cutters. Plastic pipe is a little more expensive, but you don't need any special tools. In fact, plumbing with plastic pipe is only slightly more difficult than playing with Tinkertoys. You can cut the pipes with a hacksaw, and instead of threading you simply glue them together. If you bury the plastic pipe carefully so that the sun doesn't crack it, plastic pipe is immortal.

2. Leaks always happen at the joints. Before burying the pipe, mark the joints with stakes or a pile of stones. Thus if you do

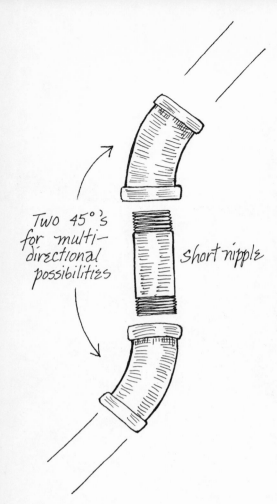

Two 45°'s for multi-directional possibilities

short nipple

manage to make a botch of the job, you won't have to dig up the whole damn pipeline to find out where you goofed.

3. Water flows downhill. Remember this, please. Meditate upon it if necessary. You'd be surprised how often people forget this simple fact when they are doing plumbing. If you are tapping into a high-pressure line, you can run your pipes any old way. But if you are dealing with gravity flow from a cistern or a spring, you have got to keep your pipes running continuously downhill. Otherwise you'll eventually create clogs, jams, and lots of troubles I wouldn't wish on anyone.

Bearing these three hints in mind, you are ready to do your own plumbing. Even if you've never done any plumbing at all, I think you can do a perfectly fine job on everything except tapping into a major water line, which should be done by someone with experience and tools.

ARRIVALS AND TRANSPLANTS

For a week or so after you've built your pond, visit it daily. It's foolish to tell you this, frankly, because for the week or so after you've built your pond, virtually nothing can keep you away. But you should be on hand to make sure that water is flowing into the pond, to catch small leaks before they get bigger, and to appreciate the marvelous things that water will do.

Within a day or two after you turn on the valve to let the water flow, your pond will have life. Insects like boatsmen, striders, and dragonflies come immediately. A few months later there will be reeds, rushes, and sedges. Within a year you can expect cattails in the pond and the seedlings of water-loving trees like willows and alders along its borders.

Cover No matter how big or small your pond is, it needs cover to protect the wildlife that will use it. Quail, rabbits, raccoons, and other small animals will appreciate a hedgerow or brush pile to protect them from predators. Try to provide at least two escape routes.

Fish   It might be a good idea to add a small, hardy native fish to your pond to help eat the mosquito larvae and eggs. In California we have a fish called a mosquito fish, and everyplace else seems to have a fish called a mosquito fish too. They may all be different species of fish for all I know, but they all do the job. Unfortunately, they often work overtime, eating the eggs and larvae of frogs, toads, turtles, and other interesting animals your pond might otherwise support. I leave you to work out the mosquito fish dilemma as best you can.

If your area has a small algae-eating minnow (you can find out from your state university), this might make a good addition to your pond.

For ponds over, say, 150 square feet, you can try out a predator fish to keep the minnows or mosquito fish under control. A bluegill is probably best. I would definitely avoid exotic fish, *especially goldfish*, which tend to take over the whole pond.

Trees   If the pond is at all sizable, the area around it will be moist enough to support some valuable trees. Your wildlife will especially appreciate any of the swamp oaks — water oak, willow oak, nuttall oak, cherry-bark oak, swamp red oak, and pin oak, all of which have nutritious and plentiful acorns.

Easy does it   What about turtles, frogs, crayfish, freshwater clams, pond lilies, and other spectacular forms of pond life? I urge you to go easy here. Be especially aware of "aquarium mania," a disease common among aquarium buffs. They no sooner get a two-quart fishbowl than they begin stocking it with every imaginable fish, plant, bubbling device, crustacean, and amphibian. The results of this disease are fatal — not to the aquarium buff, but to the creatures that get thrown willy-nilly into his fishbowl.

In short, if you go in for heavy stocking of your pond, you will have to cope with many failures and be responsible for the death of many small creatures. You'll be disappointed and you'll build up some pretty bad karma as well. So go lightly, add little, and let the water relate to your land. Visit your finished pond as you might visit your grown child: not as its creator but as its guest and admirer, willing to help out in an emergency but otherwise careful not to interfere. Appreciate what is happening, keep your senses alive, and you'll be surprised and delighted with the vigorous life that will appear spontaneously around your pond.

READING

Except for the U.S. Forest Service *Wildlife Habitat Improvement Handbook* (mentioned on page 24), with its section on rain-catching contraptions, I haven't found any really useful books on building small ponds and watering holes.

✒11 *Happy Trails to You*

Making a trail might go against your idea of what unspoiled land should be. In a sense you'll be working against nature. No matter how proud you may be of your trail, nature will treat it contemptuously. It will do everything possible to wash it away, cover it with rocks and branches, and vegetate it into obscurity.

Yet if your land has thick brush or undergrowth, you will, of course, need a trail just to get around. But even on relatively open land there's a very good reason to make a trail. In fact, if you hike frequently over your land, you're going to make a trail, whether you like it or not. You'll make a trail just as surely as a deer makes a trail, a cow makes a trail, even a mouse makes a trail. Your body will break through brush. Your big fat boots with their Vibram soles will disturb and compress the soil. Erosion will follow in your footsteps. Think about it. As long as you are destined to make a trail one way or another, you might consider making one properly and deliberately — a trail that will be pleasant to hike along and will cause minimal damage from erosion and trampled vegetation.

Another very good reason for making a trail is that it's exciting — and sometimes even fun. Especially with kids, trail building seems to connect up with the rich fantasy life of being a pioneer and explorer. It doesn't matter how small the woods happen to be.

Land without a trail is "the impenetrable jungle."

"Are we the first people ever to set foot here?" kids would ask as I'd lead them into the forest to help me build a trail.

"It could be," I'd answer, and for the rest of the day the kids would be superalert. It's almost as if they expected to stumble upon an Indian burial ground, pirate's gold, a mountain lion's den, the home of Tarzan and Jane, or perhaps even a dinosaur left over from a previous age.

Of course, no one ever found a dinosaur, and sometimes the fantasy would come crashing down with the discovery of a not-so-ancient Coke bottle. But that's not really very important. As long as the kids are alert, looking, and wondering — as long as they *feel* that they are in a brand new place — they'll turn up plenty of things.

I've had joyful times working with kids, but trail making, more than most other projects, points up one of the greatest drawbacks — namely, the quality of work you can expect. I have lots of artsy ideas about what the perfect trail should look like. It should look accidental rather than "built." It should be simple yet varied, unobtrusive yet clearly visible. I want all the scars hidden and the trees and shrubs along the sides pruned with consummate sensitivity.

These fantasies about the ideal trail are as likely to be fulfilled as a kid's fantasies about finding a dinosaur. Kids' work is bound to be sloppy and half done. That is the nature of kids. So we have a tacit agreement. The kids tolerate my artsy ideas and I tolerate their sloppiness. Nature, needless to say, tolerates everyone and sets about in good time to repair and obscure whatever aesthetic mistakes any of us make.

Planning the work There are some parts of trail making that are fun and other parts that involve hard work. I've always had a marvelous time hacking my way through the brush, following the route I had previously laid out. If there were no other considerations, I'd happily hack my way right to the end of the trail. At the end of the day I'd feel like a·member of the Explorers' Club who had reached his goal. But on the following days I'd feel more like a member of a chain gang. After you've hacked through the brush, what is left is the hard work of grading, digging, and hassling the drainage problems. My advice is to move slowly. After you've staked out the route, do one section at a time as completely as possible. That way every day you will have a taste of the hard work as well as a taste of the trailblazer's exhilaration.

Planning the trail Before you plunge into the woods with a machete in your hand and a crazy gleam in your eye, let's think about what you want in a trail.

FIRE BREAK   You might want your trail to double as a fire break or a place where crews can maintain a fire line. In some places, especially California, this is an important consideration. There are all sorts of esoteric things you must take into account in building a firebreak: prevailing winds, inflammability of different types of vegetation, conditions of the crown, availability of water, accessibility to trucks, visibility from the air, and more. Your local fire department can give you the lowdown. Listen to them eagerly, but I hope you'll keep this perspective in mind: you are building a trail first, a fire line second. In other words, do not make the perfect fire line (which is usually ugly) and hope to use it as a trail. It will make a poor trail. Instead, concentrate on making a good trail, adapting it wherever possible to its secondary function as a firebreak or potential fire line.

LOOP   Consider making a loop trail. This won't always be possible or necessary, and it will involve extra work, but it's a lot more exciting to return along a different route from the one you arrived on.

BYPASS FRAGILE PLACES   If you are expecting heavy public use of a trail, do not build it right through especially beautiful and fragile areas. Bypass them and create spur trails to serve them. This is sort of sneaky, I know. And it will add footage to your trail building. But it's the best way I know of protecting fragile areas while still making them accessible.

EASY OR HARD? If the trail is primarily for your own private use, it doesn't matter very much whether it's easy or hard. Your body will soon accustom itself to the steepness and ruggedness of any trail if you hike it often, and you'll scarcely notice the difference. If the trail is open to the public, however, I'd make it as easy as possible. I know that some backpackers intentionally seek out the toughest trails and race along them, covering as much distance as possible. For them hiking is a sort of athletic event. There is nothing wrong with this, and in fact it can be quite exhilarating. But not everyone wants to run an obstacle course. For most people a tough trail gets in the way of their enjoyment. In thinking about where to put a public trail, I would stick to places where you can maintain an easy grade.

Surveying Your first real work is to lay out the route for your trail. You know where the trail begins, you know where it ends, and you are now going to figure out the best way of connecting the beginning with the end.

Your basic criteria will be to: (1) keep the trail varied, winding, and interesting; (2) maintain a fairly steady grade (no roller coasters, please); (3) avoid traversing steep slopes; (4) stay away from loose duff and rock rubble; and (5) skirt marshes and wet spots.

Naturally, you won't be able to attain all these goals. If you *must* traverse a steep slope, go through a marsh, or cross a talus bed, I'll show you how to do it later in this chapter. But do your best to avoid these difficult situations. An extra day or two spent surveying for an easy trail route may save you a week of hard work once the trail building starts.

Depending on how much work you're willing to do, you can put a trail almost anywhere. But there is one never-to-be-broken rule you must follow: *Never Head A Trail Straight Up A Hill*! Not only is it tiring, but if a person can hike straight up, water can wash straight down and cause all sorts of nasty erosion.

The most popular method of laying out a trail is the "wandering-cow method." That is how the streets of Boston were supposedly

marsh crossing

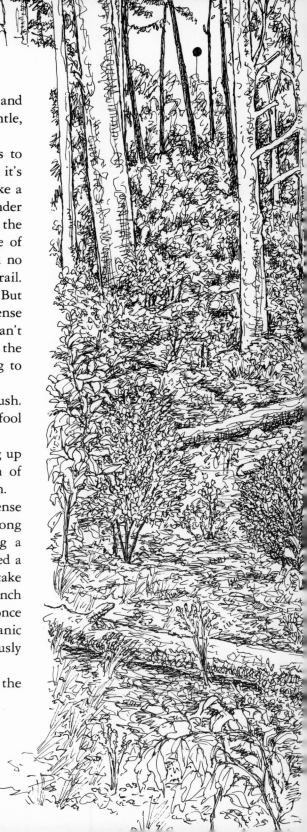

designed. The city founders let their cows wander over the land and then built the streets to follow the cow paths. The result is gentle, winding, organic streets full of interest and surprise.

If you don't have a cow of your own, or if your cow refuses to wander, you will have to take over the job yourself. Believe me, it's not at all a bad job. All you have to do is think like a cow, feel like a cow, amble like a cow. While you meditate upon clover, wander contentedly and unhurriedly from the beginning of the trail to the end. Any time you have to make a decision, choose the course of least resistance. If it seems as if you had a fairly nice walk with no impossible obstacles, *voilà*, you have just mapped out a perfect trail.

The wandering-cow method will work in almost all cases. But let's say you're building a trail through thick chaparral or dense forest. You are forever over your head in nasty bushes, you can't see more than fifty feet in any direction, and you're lost from the minute you enter the forest. No cow in her right mind is going to wander through stuff like this. At least not contentedly.

Here is one way of laying out a trail through dense forest or brush. Find a friend who will stand at the end of the trail and yell his fool head off while you walk steadily toward his voice.

Or sometimes you can go to the end of the trail and run a flag up the tallest tree to serve as a guide. Or you can float a batch of helium balloons — if you can find a convenient source of helium.

However you do it, surveying a trail through brush or dense understory can be physically hard work. Be sure to take along gloves, a long-sleeved shirt, and tough dungarees. Also bring a canteen and your favorite brush-cutting tool. If you've never used a machete, I suggest you try one. It takes a while to learn how to take care of it, keep it sharp, and use if effectively. (Try cutting a branch from below rather than hacking down at it from above.) But once you learn how to use it, you'll find a machete to be a very organic tool. You'll eventually be using it as gracefully and unconsciously as you now use a spoon or a pencil.

In surveying a trail, of course, you'll need something to mark the

trail with. Hansel and Gretel have discredited the use of bread crumbs. It's much better to use ribbons, plastic ribbons if possible, which you can tie to trees and bushes along the way. You might want to pick up two or more different colors. Use one color to mark the various experimental passes you make and the other color to mark out the final route you have chosen.

Switchbacks Before going any further, let me tell you what I think about switchbacks. As I've already said, you *never* head a trail straight up a hill. But let's say you're at the bottom of a hill; how do you get to the top? The traditional way of doing it is with a system of zigzags called switchbacks.

In theory, switchbacks are perfectly fine. They look very fancy and professional. The only problem with them is that they don't work. Ever! No sooner do you build a switchback than people begin cutting across it, creating slides and gullies straight down your hill. To counter this you can, I suppose, put up signs that say Keep on the Trail. Or you can put up educational signs that explain what terrible damage cutting across the trail does. The purpose of these signs is to change human behavior. Lots of luck! I personally think you'd be better off accepting human behavior and changing the design of your trail to eliminate switchbacks — or at least reduce their number.

One way to minimize switchbacks is to make the zigzags much longer. Instead of making six short zigzags, try doing it with only two or three more leisurely ones.

Another way to eliminate switchbacks is to make the trail extremely steep. Not straight up and down the hill, but still at an outrageous angle. You can even build steps. This will make for uncomfortably steep hiking, but over short distances it may be better than switchbacks that are doomed to fail.

Clearing I won't insult you by giving you the "standards" for how wide, how steep, how high, how many rest stops, etc. Such things depend on your taste and the usage you expect from your trail. My own preference is to make the trail narrow through open forests and meadows. In thick undergrowth or chaparral, however, I widen the trail considerably — partly to give a view, partly to give grasses and wildflowers a chance to spring up along the sides of the trail, but mostly to avoid the hassles of continually maintaining a narrow trail against the ambitions of invading brush.

Most of the actual work of clearing a trail is fairly easy. You move along rapidly. Use lopping shears and a saw to clear away the vegetation, keeping in mind the instructions for pruning. You might also remember that growth is stimulated when the tip of a branch is pointed up, and that growth tends to be suppressed when the branch is pointed down. Consequently, when trimming the shrubs along the trail's edge, I often leave more down-facing branches than I otherwise might.

Grading and sloping   After you've got the vegetation cleared away, use shovels and mattocks to clear the ground, remove rocks, and smooth the tread. As you smooth, think continually about drainage. Your trail will eventually get compacted with use, and the water, instead of being absorbed, will run off. If it runs off in many places, there won't be much damage. The problem is that sometimes the water gets channelized and washes down the trail to create a trough. In time the stream of water breaks through the trail's edge, causing all sorts of damage to the trail and to the land below. Erosion eats up trails even faster than do horses.

There are many things you might do to minimize erosion damage while you are building the trail. For example:

OUTSLOPING   By sloping the trail *gently* away from the hillside, water will be drawn off the trail continually. It will never collect, and you will solve most drainage problems before they begin.

INSLOPING   If the edge of the trail is fragile, or the slope very steep, you might have to slope the trail in toward a hillside. Water will now gather and flow along the inner bank. This water must be gotten rid of frequently; otherwise, it will become a raging torrent. The best ways of getting rid of it are with a wide,

outsloping

insloping

outsloping with water bars

rock-filled depression that cuts across the trail (a swale), or by leading it away wherever the trail makes a bend.

THANK-YOU-MA'AMS With such a beautiful name, you'd expect a thank-you-ma'am to be at least a hybrid rose. Sorry, but it's nothing more than a trench dug diagonally across the trail. It works well for an emergency, but it tends to fill in after a season or two.

WATER BARS A water bar is a log buried in the trail, usually diagonally, that diverts the water from the center of the trail to the edge.

Special problems Once you've cleared the tread and attended to the drainage, that's all there is to it — 90 per cent of the time. Unless, of course, you run into special problems like crossing marshes, crossing talus slopes, traversing very steep slopes, or having to build steps. There are drawings to show you how to handle these situations. Handle them in a deliberate, craftsmanlike manner — and if you can possibly help it, handle them as infrequently as you can. In other words, try to route the trail away from problem areas.

Clean up It's been said that great artists hide their art. Well, great trail builders hide — no, not their trails, definitely not their trails — but their trail building.

Loose earth can be thrown downhill, and the first rains will wash it away. Just don't throw it in piles, but spread it out.

The slash can be pulled off the trail and piled into brush piles for animals.

Some of the prunings can be collected and used as cuttings to make new plants.

Small trees and shrubs in the middle of the trail can be dug up and transplanted.

Disturbed soil should be treated for erosion.

And bare ground should be replanted with grass and wildflower seeds.

I hate to sound like a nag, but for goodness' sake, clean up after yourself!

Afterwards Your trail is now ready for use. You wait and watch the weather. No, you are not waiting for a perfect, clear, sunny day when you can don your *lederhosen*, put a sprig of edelweiss in your hat, and walk out whistling a happy tune. Quite the opposite. You are waiting for a cold, wretched, nasty rainstorm. When it happens, put on your rain gear, grab a shovel, and hike along the trail. Keep your nose to the ground. What is the water doing? Is it flowing off the trail in many places? Fine! You may now lift your nose and enjoy the view. If it is channelizing, however, you'll have to break up the channels right away. You can do it temporarily with a few thank-you-ma'ams, but be sure to return later to build something more permanent — like a water bar or a decent outslope.

The trail is now built. Of course, every year during the spring and summer you'll have to do some light clearing. You will be especially alert for damage after long periods of rain. But by and large your work is finished. You have built the trail, and it is now time to let the trail help build you. Go hiking often and bring friends. You're undoubtedly proud of your craftsmanship. You want to show your friends all the tricks you've learned and all the difficulties you've overcome. You point out a particularly elegant example of outsloping. But your friends aren't looking. They really don't care. Instead they are gawking at the trees, gawking at the flowers, gawking at the birds. And that's exactly as it should be. Congratulations! Your trail is a success!

READING

Believe it or not, I haven't found any good books about making trails. For that matter, I haven't found any bad books either.

With the exception of the book listed below, there doesn't seem to have been anything written on the subject.

Trail Planning and Layout, by Byron L. Ashbaugh and Raymond J. Kordish. Information Education Bulletin No. 4. New York: National Audubon Society, revised edition, 1971.

This book covers mostly self-guiding nature trails, but it has some ideas and techniques you can use for other trail building as well. Copies may be ordered from the National Audubon Society, Nature Center Planning Division, 950 Third Avenue, New York, N.Y. 10022. The price is $3.00.

talus slope:
trail built out, not cut
into slope

✎12 *Working with Kids*

This chapter is for teachers, youth leaders, and those of you who would like to get a volunteer conservation program going. Perhaps you'd like to collect a group of kids, walk out with them to where the beautiful and miraculous are spread out before you like a feast, and do some honest, helpful work. Here is how I got the conservation program at Redwood Park together, the mistakes that I made, and how I think it might be done better.

Who are your volunteers? If you're beginning a conservation program from scratch, a lot of them are going to be Boy Scouts. In fact, Boy Scouts are actively looking for conservation projects, and once they hear you have something going, your phone won't stop ringing. This is partly due to Project SOAR — Save Our American Resources. SOAR has lots of limitations. It tends to be nationalistic and sanctimonious. Yet for all its drawbacks it does succeed in dragging kids away from their TV sets and bringing them out into the woods. What happens to them after that is largely up to you.

Before going on to other volunteers, let me get the Boy Scouts off my chest. A lot of conservationists have dealt with Boy Scouts and been profoundly turned off. A troop calls in asking for a project, arrives at the proper time and in the proper uniform, does a halfhearted piece of work, and disappears. The kids get merit badges, a few trees get put into the ground, a few bags of litter get picked, but the whole experience lacks a sense of excitement or significance.

The program at Redwood Park was different, and the difference lay in this: I never took the Boy Scouts at their word. They came looking for a merit badge; I gave them an experience. I have little respect for the Boy Scouts as an organization. I dislike their militarism, their piety, their sexism, their emphasis on Achievement in the narrowest sense of the word. I didn't throw it up to them, but when it came up, I was honest about my feelings. To my surprise, many of the kids and younger Scout leaders felt the same way. Most kids do not join the Boy Scouts because they want to sing songs about keeping themselves morally clean or because they want to learn how to fold the flag properly. They join because that is the only way they can go camping, canoeing, and hiking. For them, getting a merit badge is just part of the game. Your challenge here, like everywhere else, is to get beyond the games into something immediate, moving, and real. If you keep this in mind, you'll stand a good chance of turning the merit badge silliness into a real human experience.

By and large I had the most fun with school kids, especially high school kids from loosely structured schools. You'll have to go looking for them and deal with school administrators to spring them out during school hours. In the San Francisco Bay area this isn't any problem. Many of the schools I worked with provided transportation to the park and gave the kids course credit in natural science or physical education for stomping around the woods.

In dealing with schools, be sure you get volunteers, not conscripts. Avoid whole classes that have come merely because the teacher told them to. Before I learned this lesson thoroughly, I often found myself facing thirty-six hostile teen-agers who had just slunk out of a school bus. In the immortal words of one teacher, I was expected to "put these damn kids to work and teach them something about the ecology." But no one ever told me how. Then one day, as I found myself looking into thirty-six teen-age smirks, I knew exactly what I wanted to do. I did what any other gentle, basically sane naturalist would have done in such a situation. With a loud and joyful "whoopee," I turned my back and ran away.

"C'mon," I yelled. "The ecology lecture is this way." And I took off like a madman — up slopes and down slopes, across meadows and through forests. Most of the kids followed me close behind, tripping, rolling, stumbling, shouting, and — best of all — laughing. We left the teachers far behind. We splashed through the stream and drank water from a fresh brook. We ate miner's lettuce underneath a big oak tree. We talked and joked, and when we got back to the bus I asked the kids, "Did you learn anything about the ecology?"

"Yeah," said one kid. "It's fun." Since that day, whenever I found myself with a group of conscripts, I did my best to have fun and left the productive work for another day.

Another place to look for volunteers is among 4-H Clubs, Girl Scouts, Brownies, Cub Scouts, YMCA, and fraternal organizations like DeMolay. What I said about the Boy Scouts applies equally here. If you have your head together, most of these groups will be more than glad to forgo their own rituals and join you on whatever your trip happens to be.

In addition to volunteer labor, I had terrific success in getting volunteer expertise. The Soil Conservation District and my local agricultural extension agent proved invaluable. I was also lucky in that several universities and colleges are within an hour's drive of Redwood Park. By knocking at doors I got the foremost experts in the country to advise me on wildlife problems, watershed management, erosion-control engineering, meadow-land ecology, and so on.

When I first came to Redwood Park there was already a conservation program, which I stepped into. One of my first acts was to beef it up with a lot of recruiting. At the beginning I attended Boy Scout meetings. The Boy Scouts of the Bay area are divided into districts, and the scoutmasters of each district meet every month or so. The main purpose of these meetings, as near as I could figure out, was to teach the scoutmasters how to tie bowlines. I found myself welcome, I suspect even refreshing, and I'd give a little speech about our conservation program. I plied the scoutmaster circuit for about two months, and it paid off with lots of human contact.

I gave the same "scoutmaster speech" to school administrators and other youth leaders. I once tried writing it out in a letter I could mail, but the letter didn't get anywhere near the results I got from knocking at doors and laying the whole thing out on someone's unsuspecting head. Throughout these speeches I kept emphasizing what I felt was the most important thing about the program: it was not designed to get free labor out of kids but to give them an educational experience.

There's something important I'd like to add about the spirit of recruiting. It has to be honest. The temptation to exaggerate is enormous. I was offering an interesting, for many kids an enjoyable, day in the woods. Sometimes, though, instead of "interesting," I found myself substituting "fascinating." For "enjoyable" I would talk about "ecstatic." Instead of describing how we planted trees or collected seeds, I found myself muttering nonsense about Saving the Ecology. Why I did it, I don't know. Afterwards, I would feel just terrible, like an encyclopedia salesman, perhaps. I've worked with groups that were promised ecstasy, enlightenment, and The Salvation of Mother Earth. We had an intensely disappointing time. On the other hand, groups who were told frankly, "We'll walk around a bit, look at how plants grow and how water flows, do a couple of hour's work, get to know each other, and maybe we'll really hit it off" — these groups tended to enjoy themselves much more. The moral is obvious: recruit, but recruit honestly. To the extent that we represent the beginnings of a new consciousness, unless we're honest we've lost it all, right at the start.

Once you get your volunteers lined up, you have to figure out the best project for them. Throughout the book I've described many of my favorite projects and the sorts of kids who seem to enjoy them best. But let me repeat something I've said before: DO NOT MAKE THE PROJECTS TOO DIFFICULT. A project cannot be too hard for them, either physically or conceptually. When I first took the conservation job at Redwood Park, I used to talk a lot about native California grasses. These were perennial bunch grasses which, under the pressures of grazing and drought, gave way to the European annuals that now dominate the grasslands. I'd explain how the annuals have failed to hold the land against invading brush, especially when fire was also controlled. I'd then discuss the ecology of chaparral and the limitations of coyote brush, the influence of gopher erosion, the adjustment of wildlife, and the theory of fire-climax ecology. Then I'd explain my scheme for restoring the native perennial grasses, and off we'd march, me and a group of open-mouthed kids who didn't have the vaguest idea of what I was talking about, and for whom this conservation project, like the rest of their lives, was a profoundly muddled piece of adult mystification that demanded bodily discomfort.

In short, I have a lot of fairly sophisticated land-management ideas and some ambitious projects in mind that I was never able to do because they were beyond the comprehension of most of my volunteers. For most kids, grass is grass, and that's all there is to it.

I've also learned to avoid projects that are physically too difficult, like pick-and-shovel work. There's no way in the world you can convince a group of kids that digging a trench is fun, educational, or rewarding.

When it comes to difficulty, by the way, don't trust the kids' estimates of what they can handle. I found that many kids have rather grandiose ideas, and they'd come to me with huge, earth-moving projects. They seemed almost relieved when I would reduce their schemes to more manageable proportions. I found it best to be firm at the beginning and then to give the group a free hand later on.

Sometimes hard physical work is unavoidable, like certain erosion-control projects, or burying a pipeline to a pond. In such cases I often adopted a policy of outright, up-front bribery. I'd choose a group of the biggest kids I could find and explain that this was a terrible project but that the work simply had to get done. In return, I'd promise something extra special for the group. I usually gave them free camping privileges and trucked them off to a swimming hole at the end of the day.

Whether a project was easy or hard, one thing I did offer was a humane experience. When kids came to the park, I did not want it to be "just another day" in their lives. I feel very sorry for the busloads of city kids who get dragged out into the woods by their teachers, lectured at, told facts, and quizzed the next day. I also feel sorry for the large numbers of alert kids for whom "ecology" no longer means the relationship between butterflies and milkweed but has come to mean poison, pollution, and catastrophe. My own approach with kids was to stress basic sensual and physical experiences, to swing on rope swings and get turned on by water, to spend as much time as we could in immediate sensations that go beyond interpretation.

I tried to give a lot to the kids who came to the park, and the main thing I had to offer was myself. I know how egotistical and platitudinous that sounds, but since it's true I'll say it anyway. I remember when I was a kid how mysterious the adult world seemed, and how grateful and amazed I was when any adult stepped out of his or her role as teacher, mailman, uncle, or whatever and made authentic human contact with me. That's what I tried to do. I avoided the stereotyped nature walks, and instead I offered kids intimacy and a frank sharing of my own sensibilities. I did not merely say, "This is a live oak." I tried to tell them something about what I was seeing and feeling in the presence of that live oak. Many kids would respond to this approach with an openness about their own feelings. Often we'd make real contact, become friends, and the kids would return again and again. During my three years at Redwood Park I suspect it was these bonds of friendship more than anything else that made the conservation program so alive and so much fun.

READING

Nature Study for Conservation, by John W. Brainerd. New York: Macmillan, 1971.

This book was sponsored by the American Nature Study Society and has an introduction by Roger Tory Peterson. It gives lots of things that kids might study, like mapping, seed experiments, a wildlife census, vegetative surveys, etc. On the surface it is a conventional, textbookish presentation about how to introduce kids to nature. But beyond that it is an excellent, sensitive book by a man who has obviously had lots of experience teaching kids — and who has learned from his experience.

Working with Nature, by John Brainerd. New York: Oxford University Press, 1973.

This second book by John Brainerd passes from merely studying nature to working with it — either with kids or by yourself. The book seems to suffer from being overorganized — both in the way the material is presented and in its general feeling toward land. Yet there are some very worthwhile suggestions on various aspects of land management, water management, vegetative management, and wildlife management. There's even a unique chapter entitled "Rock Management," with instructions for removing rock from outcroppings, breaking it up, moving boulders, and building stone fences.

Further Help

There are two potentially valuable sources of advice and expertise — your local agricultural extension agent (often listed as the county farm and home adviser) and the Soil Conservation District (an agency of the U.S. Department of Agriculture). I say *potentially* because what you get from either agency depends largely on who's running it in your area.

Agricultural extension agents, like doctors, seem more and more to be getting away from home visits. But they are accessible by telephone. In fact, all day long their telephones are ringing with calls from people who want to know why their lettuce is wilting, why their roses are dropping off, why their lawns are being attacked by brown spots, and how to can peaches, build nesting boxes for wrens, or plant avocado pits. Agricultural extension agents are expected to be omniscient, and needless to say, they are not. Competent and helpful, if you're lucky, but definitely not omniscient. If you feel that your local agent is not answering your question very well, ask him to refer you to someone else — a state forester, a fish and wildlife agent, or perhaps a local nursery operator.

The Soil Conservation District, on the other hand, is far more likely to send someone out to your land — as long as you can convince them that what you are doing is in some way related to conservation of soil. In the areas of pond building or erosion control, they may even supply you with free engineering advice.

A possible source of money for your projects is the Agricultural Stabilization and Conservation Agency, also part of the U.S. Department of Agriculture. This is the agency that for the last few decades has been paying farmers to keep land idle and keep food off the market, thus "stabilizing" the economy. Less well known are its conservation functions. For certain projects like reforestation, erosion control, and water conservation, there is a cost-sharing plan in which the government foots part of the bill. The catch is that if your project is approved, it must be executed according to strict standards: you may find out that the forest you had in mind is considerably different from what the agency had in mind. Nevertheless, if you have a large piece of land on which you are planning to do a lot of work, by all means get in touch with your local agent to see where you agree and what you can work out.

❧Index

Acorns: as mast, 21-22; stratification of, 99; *See also* Nuts; Nut trees; Oaks
Agricultural extension agent, 182
Agricultural Stabilization and Conservation Agency, 183
Alder, 20, 129
Alfalfa, 24
Apple tree, 22, 142
Ash, 22, 142
Aspen, 20

B

Barbed wire, 54-56
Bare root: seedlings, 127-34; transplants, 147
Barley, 71
Bayberry, 16
Bears, 9
Beech, 57
Bentonite, 159
Birch, 20
Bittersweet, 10
Blackberry, 10, 108
Boy Scouts, 127, 175-76, 178. *See also* Kids

Brush. *See* Chaparral
Brush mats, 76, 78
Brush piles, 9-12; brambles serve as, 128; from a felled tree, 34; for fish, 10; from layered plant, 110; near ponds, 162-63
Brush wattles, 64, 77-78
Buckeye, 142
Buckthorn, 16
Buckwheat hulls, 45
Burns: cause erosion, 66; help maintain openings, 19; affect wildlife, 9
Butternut, 21. *See also* Nuts; Nut trees

C

Callus, 52, 115-16
Catalpa, 114, 142
Cats (feral), 17-19
Ceanothus, 16
Cement (for tree surgery), 52
Chaparral: corridors through, 17; replaces grassland, 68, 179; and reforestation, 128; trails through, 171
Check dams: in gullies, 79-83; hy-